Network Design
for IP Convergence

Network Design for IP Convergence

Yezid Donoso

CRC Press
Taylor & Francis Group
Boca Raton   London   New York

CRC Press is an imprint of the
Taylor & Francis Group, an **informa** business
AN AUERBACH BOOK

Auerbach Publications
Taylor & Francis Group
6000 Broken Sound Parkway NW, Suite 300
Boca Raton, FL 33487-2742

© 2009 by Taylor & Francis Group, LLC
Auerbach is an imprint of Taylor & Francis Group, an Informa business

Library of Congress Cataloging-in-Publication Data

Donoso, Yezid.
  Network design for IP convergence / Yezid Donoso.
    p. cm.
  Includes bibliographical references and index.
  ISBN 978-1-4200-6750-7 (alk. paper)
    1. Computer network architectures. 2. Convergence (Telecommunication) 3. TCP/IP (Computer network protocol) I. Donoso, Yezid. II. Title.

TK5105.52.D66 2009
004.6'5--dc22   2008043273

Visit the Taylor & Francis Web site at
http://www.taylorandfrancis.com

and the Auerbach Web site at
http://www.auerbach-publications.com

To my wife, Adriana—

for her love and for our future together.

To my children, Andres Felipe, Daniella, and Marianna—

a gift of God to my life.

Contents

x ■ *Contents*

Preface

The need to integrate services under a single network infrastructure in the Internet is increasingly evident. The foregoing is what has been defined as convergence of services, which has been widely explained for many years. However, in practice, implementation of convergence has not been easy due to multiple factors, among them the integration of different layer 1 and layer 2 platforms and the integration of different ways of implementing the concepts of quality of service (QoS) under these technological platforms.

It is precisely for the abovementioned reasons that this book was written; it aims to provide readers with a comprehensive, global vision of service convergence and especially of IP networks. We say this vision is "global" because it addresses different layers of the reference models and different technological platforms in order to integrate them as occurs in the real world of carrier networks. This book is comprehensive because it explains designs, equipment, addressing, QoS policies, and integration of services, among other subjects, to understand why a specific layer or a technology may cause a critical service to not operate correctly.

This book addresses the appropriate designs for traditional and critical services in LAN networks and in carrier networks, whether MAN or WAN. Once the appropriate design for these networks under the existing different technological platforms has been explained in detail, we also explain under the multilayer scheme the concepts and applicability of the QoS parameters. Finally, once infrastructure has been covered, we explain integration of the services, in "not real time" and "real time," to show that they can coexist under the same IP network.

The book's structure is as follows:

Chapter 1—In Chapter 1 we explain some basic concepts of networks, the layer in which some of the most representative technologies are operating, and operation of some basic network equipment.

Chapter 2—This chapter concentrates on the specification of a design that's appropriate for a LAN network in which converged services are desired.

Chapter 3—Chapter 3 specifies a design that is appropriate both in the backbone and in the last mile in MAN and WAN carrier networks, in order to appropriately support service convergence.

Chapter 4—Chapter 4 introduces the different QoS schemes under different platforms and explains how to specify them for critical services in order to successfully execute service convergence.

Chapter 5—Finally, Chapter 5 discusses service convergence for not real-time and real-time applications, and how these services integrate to the carrier LAN and MAN or WAN network designs.

About the Author

Yezid Donoso, PhD, is currently a professor of computer networks in the Computing and System Engineering Department at the Universidad de los Andes in Bogotá, Colombia. He is a consultant in computer network and optimization for Colombian industries. He has a degree in system and computer engineering from the Universidad del Norte (Barranquilla, Colombia, 1996), an MSc degree in system and computer engineering from the Universidad de los Andes (Bogotá, Colombia, 1998), a DEA in information technology from Girona University (Girona, Spain, 2002), and a PhD (cum laude) in information technology from Girona University (Girona, Spain, 2005). He is a senior member of IEEE and a distinguished visiting professor. His biography is published in the following books: *Who's Who in the World*, 2006 edition; *Who's Who in Science and Engineering* by Marquis Who's Who in the World; and *2000 Outstanding Intellectuals of the 21st Century* by the International Biographical Centre, Cambridge, England, 2006. He received the title Distinguished Professor from the Universidad del Norte (Colombia, October 2004) and a National Award of Operations from the Colombian Society of Operations Research (2004). He is the co-author of the book *Multi-Objective Optimization in Computer Networks Using Metaheuristics* (2007). He can be reached via e-mail at ydonoso@uniandes.edu.co.

List of Translations

Spanish	English
Número de canales de enlaces de VoIP	Number of VoIP channels
Número de canales de abonados de VoIP	Number of VoIP trunking
Calidad de servicio IP	IP quality of service
Extensión	Extension
Plan numeración Pal	Main numeration plan
Plan de numeración público	Public numeration plan
Plan de numeración privado	Private numeration plan
Plan de numeración público restringido	Restricted public numeration plan
Añadir	Add
Borrar	Delete
Modificar	Update
Cancelar	Cancel
Placa IP	IP values
Placas	Values
Direcciones IP para PPP	IP address for PPP
CPU principal	Main CPU
Acceso internet	Internet access
VoIP (esclavo)	VoIP (slave)

Spanish	English
Dirección de router por defecto	Default gateway address
Máscara subred IP	IP subset mask
Ayuda	Help
Direccion IP	IP address
Nombre gateway VoIP	VoIP gateway name
Configuración IP de Windows 2000	Windows 2000 IP configuration
Nombre del host	Host name
Sufijo DNS principal	Main DNS suffix
Tipo de nodo	Node type
Enrutamiento de IP habilitado	IP routing active
Proxy de wins habilitado	Wins proxy active
Ethernet adaptador conexión de área local	Ethernet adapter local area connection
Descripción	Description
Dirección física	Hardware address
Mascara de subred	Subnet mask
Puerta de enlace predeterminada	Default gateway
Servidores DNS	DNS servers
Fichero de datos	Date file
Cliente	Client
Tabla ARS	ARS table
Inicio	Start
Herramientas	Tools
Proveedor	Provider
Instalación típica	Typical installation
Modificación tipica	Typical update
Numeración	Numeration

Spanish	English
Números de instalación	Installation numbers
Configuración por defecto	Default configuration
Plan de numeración	Numeration plan
Códigos de servicio	Service codes
Tabla modif. números DDI	Update table DDI numbers
Tabla prefijos fraccionamiento	Fraction prefix table
Fin de marcación	Dial end
Selección automática de rutas	Automatic route selection
Tabla ARS	ARS table
Lista de grupos	Group list
Tabla de horas	Hour table
Grupos del día	Day groups
Operadores/destinos	Operators/destination
Códigos de autorización	Authorization codes
Tono/pausa	Tone/pause
Varios ARS	ARS various
Conversion PTN	PTN conversion

Chapter 1

Computer Network Concepts

This book addresses network design and convergence of IP services. In this first chapter, we present a basic analysis of computer network fundamentals. For more in-depth information on these topics, we recommend reading [KUR07], among others.

In this first chapter we present fundamentals on the main difference between digital transmission and analog transmission, network classification according to size, some network architectures and technologies, and basic functions that take place in computer networks; last, we explain the operation of the different devices with which network design will subsequently take place.

1.1 Digital versus Analog Transmission

When transmitting over a computer network one can identify two concepts. The first concept is the nature of the data, and the second is the nature of the signal over which such data will be transmitted.

The nature of data can be analog or digital.

An example of digital data is a text file forwarded through a PC. The characters of this file are coded, for example, through the ASCII code, and converted to binary values (1s and 0s) called bits, which illustrate said ASCII character (Figure 1.1). The same would happen for a binary file (.doc, .xls, .jpg, etc.), in which binary value illustrates part of the information found in the file, whether through a text processor, spreadsheet, or images.

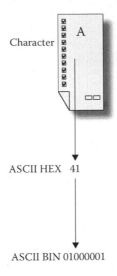

Character   A

ASCII HEX   41

ASCII BIN 01000001

Figure 1.1   Digital data.

An example of analog data is transmitted voice or video. In voice the information is analog; here we produce sound, where its tones are represented as continuous waves over time and do not display discrete specific values, as happens with digital data. Figure 1.2 shows the transmission of analog data of a person using the phone.

The nature of the signal can also be analog or digital.

An example of digital signals is transmission of information from a PC to the computer network (Figure 1.3). Here, the PC produces digital information (consisting of a series of 1s and 0s called bits) that will be transmitted over the data network also as 1s and 0s. This does not mean that a direct relationship exists between the physical way in which the PC transmits a 1 and how the computer network understands a 1; in other words, there are different kinds of coding schemes to

Figure 1.2   Analog data.

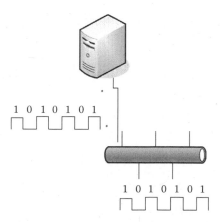

Figure 1.3   Digital signals.

represent bits and these can vary from PC to network and even between different network technologies. For example, in transmissions over copper wires there are network technologies that understand a 1 as a high-voltage (+5 V) level and other technologies may understand the 1 as –5 V. It is important, therefore, to correctly interconnect the physical transmission of these technologies so that when a transmitting device forwards a bit as 1 and +5 V, the receiver understands that +5 V means a 1 at the bit level.

An analog signal is data or information transmitted as a continuous signal regardless of the nature of the data. Figure 1.4 shows that a PC is transmitting digital data over analog signals.

We have seen that to transmit information one can use digital or analog signals. The question now is how to represent an analog signal and how to represent a digital signal.

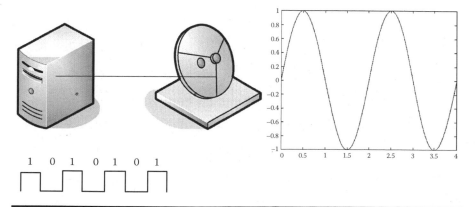

Figure 1.4   Analog signals.

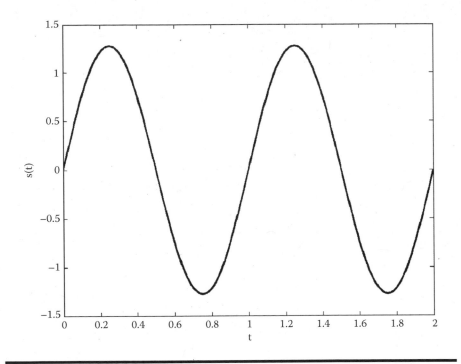

Figure 1.5   Sin function.

The main characteristic of analog signals is their continuous form with small changes in the value of the function. Analog signals can be represented through Fourier series, and in this case every analog transmission could be represented by a combination of sinusoidal or cosinusoidal functions. In this case we could say that we have a function that can be described in the following way:

$$s(t) = \sin(2\pi f t)$$

Here we can say that the value of the signal in the time (t) domain depends on the frequency (f) being used; we would thus be illustrating an analog signal, which is continuous in time. To show an example, we could say that $T = 1$ where $T = 1 / f$ and realizing t from 0 to 2 with 0.01 increments; this 0.01 value has been randomly selected for this example and any real value could be used to better represent the curve of the function. Figure 1.5 shows the signal's behavior for these example values. We can see in this figure that the behavior is that of a continuous signal through which we could represent some kind of data.

We could combine a set of sinusoidal functions in such a way that the frequencies used are in the range from 1 to ∞ and using odd values. The odd values of the sin function are used so that they don't counteract the values of the function. We could also make every signal generated by a frequency (f) let a multiple of

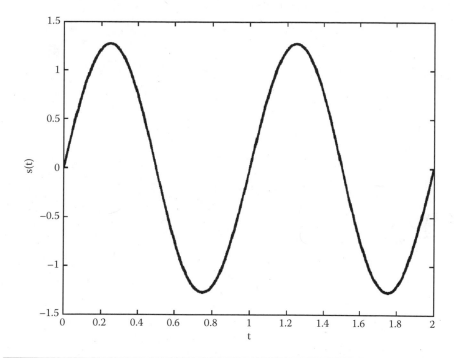

Figure 1.6   *s(t)* function with *n* = 1.

the fundamental, that is, of (1*f*); and also that the value of *s(t)*, the signal as we increase the frequency, (*kf*), be smaller every time due to the factor K inside the Sin function.

The following equation could describe this.

$$s(t) = \frac{4}{\pi} \sum_{k=1, k\ odd}^{n} \frac{\sin(2\pi kft)}{k}$$

One can see that every new signal generated by a frequency (*kf*) multiple of the fundamental (1*f*) generates an oscillation within the fundamental frequency. Since the value of the oscillation within the total is also being divided by *k*, this means that as the number of frequencies (*k*) increases, its incidence over the value of the resulting signal [*s(t)*] is increasingly smaller.

The following scenarios result from the foregoing equation by changing the value of *k*, that is, the number of frequencies used.

For all cases *T* = 1. We will begin by showing the effect when *n* = 1 (Figure 1.6), that is, when we only have frequency 1*f*. In this case the signal is similar to Figure 1.5.

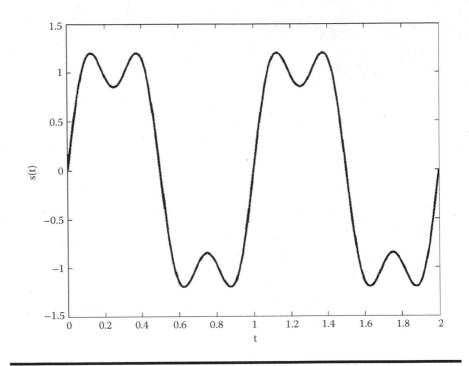

Figure 1.7 *s(t)* function with *n* = 3.

In this new case we show the effect when *n* = 3 (Figure 1.7). In other words, we have frequencies 1*f* and 3*f*. The resulting effect is that the signal generated by 3*f* superimposes the signal generated by 1*f*, with the additional behavior that the most influential signal over the value of *s(t)* is 1*f*, since the oscillation of 3*f* is much smaller (one-third) than that of 1*f*.

The next example shows the resulting signal behavior [*s(t)*] when *n* = 5 (Figure 1.8). It is evident in this figure that there are more oscillations (3*f* and 5*f*) in terms of the fundamental frequency (1*f*).

We could successively continue trying with different values of *n* and every time we will find more oscillations in terms of the fundamental frequency (1*f*). For example, if we have *n* = 19 (Figure 1.9), the signal will increasingly look more like a digital style signal. The important thing is that the electronic device can understand during the time of a bit if the value is, for example, in this case, +1 or −1. We can conclude from the foregoing that more or fewer frequencies are used to represent a digital signal.

Finally, if we use *n* = 299 (Figure 1.10) in this example we see a good representation of a digital signal, which in this case has only two values, +1 or −1. The example that we have used does not intend to tell us that electronic devices must necessarily use this range of frequencies; it has been used for educational purposes.

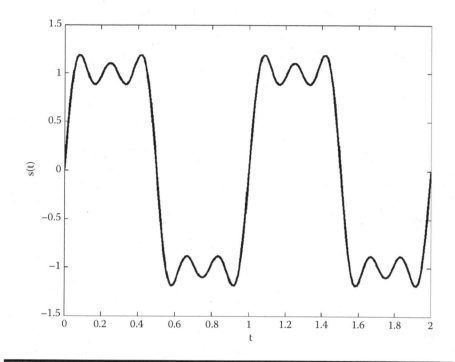

Figure 1.8 *s(t)* function with *n* = 5.

1.2 Computer Networks According to Size

Computer networks may be classified in several ways according to the context of the study being conducted. The following is a classification by size. Later on we will specify to which of the networks we will apply the concepts of this book.

1.2.1 Personal Area Networks (PANs)

Personal Area Networks are small home computer networks. They generally connect home computers to share other devices such as printers, stereo equipment, etc. Technologies such as Bluetooth are included in PAN networks.

A typical example of a PAN network is a connection to the Internet through the cellular network. Here, the PC is connected via Bluetooth to the cell phone, and through this cell phone we connect to the Internet, as illustrated in Figure 1.11.

1.2.2 Local Area Networks (LANs)

Local Area Networks are the networks that generally connect businesses, public institutions, libraries, universities, etc., to share services and resources such as the

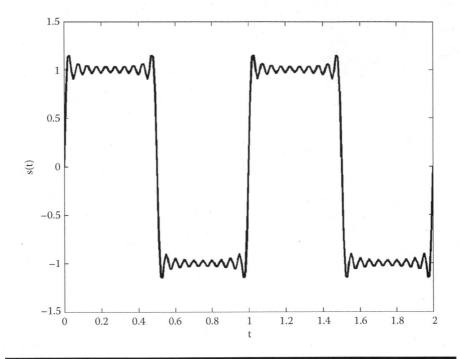

Figure 1.9 *s(t)* function with *n* = 19.

Internet, databases, printers, etc. They include technologies such as Ethernet (in any of its speeds, today reaching up to 10 Gbps), Token Ring, 100VG-AnyLAN, etc.

The main structure of a LAN network traditionally consists of a switch to which the switches of office PCs are connected. Corporate servers and other main equipment are also connected to the main switch. Traditionally, this main switch, which can be a third layer switch or through a router connected to the main switch, is the one that connects the LAN network to the Internet. This connection from the LAN network to the carrier or ISP is called the *last mile*.

Figure 1.12 shows a traditional LAN network design.

1.2.3 Metropolitan Area Networks (MANs)

Metropolitan Area Networks are networks that cover the geographical area of a city, interconnecting, for instance, different offices of an organization that are within the perimeter of the same city. Within these networks one finds technologies such as ATM, Frame Relay, and xDSL, cable modem, RDSI, and even Ethernet.

A MAN network can be used to connect different LAN networks, whether with each other or with a WAN such as Internet. LAN networks connect to MAN networks with what is called the *last mile* through technologies such as ATM/SDH, ATM/SONET, Frame Relay/xDSL, ATM/T1, ATM/E1, Frame Relay/T1,

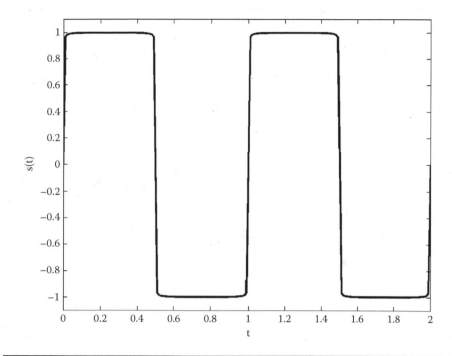

Figure 1.10 *s(t)* function with *n* = 299.

Frame Relay/E1, ATM/ADSL, Ethernet, etc. Traditionally, the metropolitan core is made up of high-velocity switches, such as ATM switchboards over an SDH ring or SONET or Metro Ethernet. The new technological platforms establish that MAN or WAN rings can work over DWDM and can go from the current 10 Gbps to transmission velocities of 1.3 Tbps or higher. These high-velocity switches can also be layer 3 equipments and, therefore, may perform routing.

Figure 1.13 shows a traditional MAN network design.

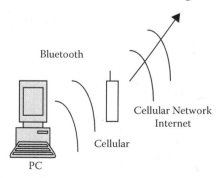

Figure 1.11 PAN design.

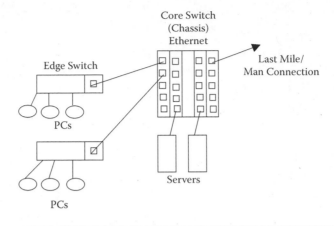

Figure 1.12   LAN design.

1.2.4 Wide Area Networks (WANs)

Wide Area Networks are networks that span a wide geographical area. They typically connect several local or metropolitan area networks, providing connectivity to devices located in different cities or countries. Technologies applied to these networks are the same as those applied to MAN networks, but in this case a larger geographical area is spanned, and, therefore, a larger number of devices and greater complexity in the analysis that must be done to develop the optimization process are needed.

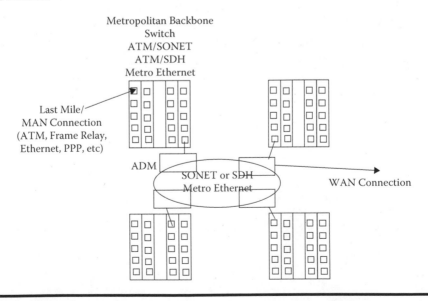

Figure 1.13   MAN design.

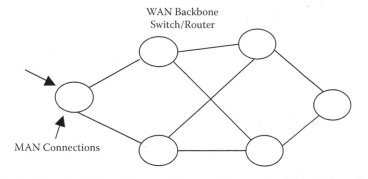

Figure 1.14 WAN design.

The most familiar case of WAN networks is the Internet as it connects many networks worldwide. A WAN network design may consist of a combination of layer 2 (switches) or layer 3 (routers) equipment, and the analysis depends exclusively on the layer under consideration. Traditionally, in the case of this type of network, what's normal is that it be analyzed under a layer 3 perspective.

Figure 1.14 shows a traditional WAN network design.

In this book we will work with several kinds of networks. Chapter 3 discusses MAN and WAN network designs.

1.3 Network Architectures and Technologies

This section presents basic concepts of some architectures such as OSI and TCP/IP, and some models of real technologies. The purpose is to identify the functions performed by each of the technologies as compared to the OSI reference model.

1.3.1 OSI

This model was developed by the International Organization for Standardization, ISO, for international standardization of protocols used at various layers. The model uses well-defined descriptive layers that specify what happens at each stage of data processing during transmission. It is important to note that this model is not a network architecture since it does not specify the exact services and protocols that will be used in each layer.

The OSI model is a seven-layer model:

Physical layer—The physical layer is responsible for transmitting bits over a physical medium. It provides services at the data link layer, receiving the blocks the latter generates when emitting messages or providing the bits chains when receiving information. At this layer it is important to define the type of physical medium to be used, the type of interfaces between the medium and the device, and the signaling scheme.

Data Link layer—The data link layer is responsible for transferring data between the ends of a physical link. It must also detect errors, create blocks made up of bits, called *frames*, and control data flow to reduce congestion. The data link layer must also correct problems resulting from damaged, lost, or duplicate frames. The main function of this layer is switching.

Network layer—The network layer provides the means for connecting and delivering data from one end to another. It also controls interoperability problems between intermediate networks. The main function performed at this layer is routing.

Transport layer—The transport layer receives data from the upper layers, divides it into smaller units if necessary, transfers it to the network layer, and ensures that all the information arrives correctly at the other end. Connection between two applications located in different machines takes place at this layer, for example, customer-server connections through application logical ports.

Session layer—This layer provides services when two users establish connections. Such services include dialog control, token management (prevents two sessions from trying to perform the same operation simultaneously), and synchronization.

Presentation layer—The presentation layer takes care of syntaxes and semantics of transmitted data. It encodes and compresses messages for electronic transmission. For example, one can differentiate a device that works with ASCII coding and one that works with BCD, even though in each case the information being transmitted is identical.

Application layer—Protocols of applications commonly used in computer networks are defined in the application layer. Applications found in this layer include Internet surfing applications (HTTP), file transfer (FTP), voice over networks (VoIP), videoconferences, etc.

These seven layers are depicted in Figure 1.15.

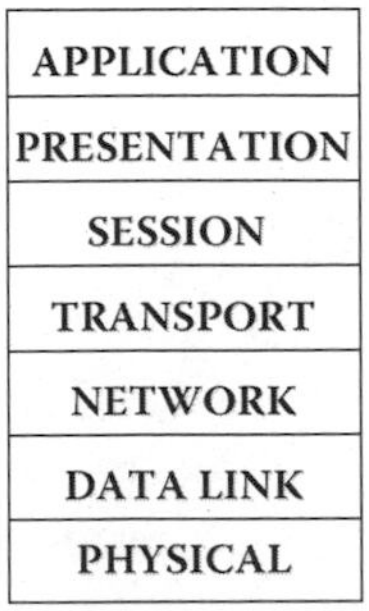

| APPLICATION |
| PRESENTATION |
| SESSION |
| TRANSPORT |
| NETWORK |
| DATA LINK |
| PHYSICAL |

Figure 1.15 OSI architecture.

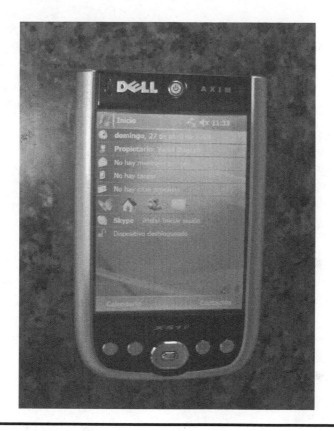

Figure 1.16 PDA.

1.3.2 PAN

PAN technologies include Bluetooth, which is described next.

1.3.2.1 Bluetooth

Bluetooth is the name of the technology specified by standard IEEE 802.15.1, whose goal traditionally focuses on allowing short-distance communication between fixed and mobile devices. Devices that most normally use this technology are PDAs (personal digital assistants) (Figure 1.16), printers, laptops, cellular phones, and keyboards, among others.

It is somewhat complicated to write with a pencil, for instance, on a PDA. For this reason, external keyboards that connect via Bluetooth to the PDA have been created, thus avoiding the use of connecting cables. In other words, it is very practical for a PDA to have Bluetooth technology so that it can interconnect to different external devices. The use of Bluetooth is no longer as frequent in laptops, since laptops now feature built-in WiFi technology that enables their connection to the

Internet, among PCs, to a computer network, etc., faster and with a wider range than with Bluetooth. The market is also seeing devices with built-in WiMAX; this means that they connect to the Internet not only at a hot spot but anywhere in the city, even on the go, with standard IEEE 802.16e. Bluetooth, therefore, is an excellent technology for devices that require little transmission Mbps and short range.

In Bluetooth version 1.0 one can transmit up to 1 Mbps gross transmission rate and an effective transmission rate of around 720 Kbps. It can reach a distance of approximately 10 m, although it could reach a greater range with more battery use due to the power needed to reach such distances with a good transfer rate. It uses a frequency range of 2.4 GHz to 2.48 GHz. Version 1.2 improved the overlapping between Bluetooth and WiFi in the 2.4 GHz frequencies, allowing them to work continuously without interference; safety, as well as quality of transmission, was also improved. Last, version 2.0 improves the transmission rate, achieving 3 Mbps, and has certain improvements over version 1.2 that correct some failures.

When compared to OSI architecture, Bluetooth is specified in layer 1 and part of layer 2, as shown in Figure 1.17.

The power range of antennae, from 0 dBm to 20 dBm, is defined in the RF (radio frequency) sublayer, associated to the physical layer. The frequency is found in the 2.4 GHz range.

The physical link between network devices connected in an ad hoc scheme, in other words, among all without an access point, is established in the baseband sublayer, also associated to the physical layer.

Bluetooth layer 2 is formed by the link manager and the Logical Link Control and Adaptation Protocol (L2CAP), which provide the mechanisms to establish connection-oriented or nonoriented services in the link and perform part of the functions defined in layer 2 of the OSI model.

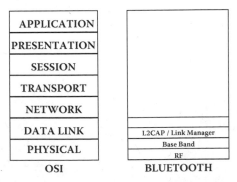

Figure 1.17   Bluetooth architecture.

1.3.3 LAN

There are many different types of LAN technologies: some are no longer used and others are used more than others. In this section we discuss only two of these technologies. Ethernet is the most common wired LAN solution and WiFi is the most widespread as wireless LAN.

1.3.3.1 Ethernet

Ethernet technology was defined by a group of networking companies and later standardized by IEEE. Layer 1 of Ethernet defines the electrical or optical characteristics for transmission, as well as the transmission rate. Layer 2 Ethernet consists of two IEEE standards. The first, standard 802.3, traditionally works with the CSMA/CD (Carrier Sense Medium Access with Collision Detection) protocol to access the medium and transmit. The second, 802.2, defines the characteristics of transmission at the link in case it is connection oriented or nonoriented. Figure 1.18 shows the layers of Ethernet technology.

1.3.3.2 WiFi

WiFi (Wireless Fidelity) technology is a set of IEEE 802.11 standards for wireless networks. Standard 802.11 defines the functions of both layer 1 and layer 2 in comparison with the OSI reference model. At the physical layer, the frequency and the transmission rate depend on the standard being used. For example, standard 802.11a uses the 5 GHz range and can transmit at a range of 54 Mbps; standard 802.11b uses the 2.4 GHz range and can transmit at a rate of 11 Mbps; and standard 802.11g uses the same range as 802.11b and can transmit up to 54 Mbps. A new standard currently being developed, 802.11n, is aimed at transmitting around 300 Mbps in the 2.4 GHz frequencies. This technology will reach distances of 30 m to 100 m depending on the obstacles, the power of antennae, and the access points.

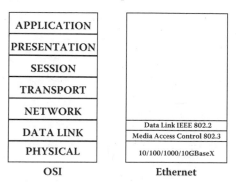

Figure 1.18 Ethernet architecture.

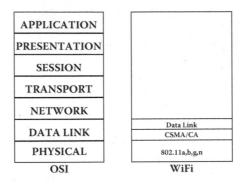

Figure 1.19 WiFi architecture.

As to layer 2 of the OSI model, WiFi defines a standard to access the CSMA/CA (Carrier Sense Medium Access with Collision Avoid) medium. Figure 1.19 shows the layers of WiFi technology.

As we have seen, when we talk about the name of LAN technologies we refer to those that perform the layer 1 and layer 2 functions as compared to the OSI model. It will therefore be necessary to go over other technologies to see how they complement with regard to network interconnection and other layers of the OSI model.

1.3.4 MAN/WAN

This section analyzes the architectures of LAN network accesses to MAN or WAN networks, commercially known as *last mile* or *last kilometer*, and the architectures that are part of operator backbones. We will find here that some technologies define only layer 1 functions and others only layer 2 functions.

1.3.4.1 TDM (T1, T3, E1, E3, SONET, SDH)

Time Division Multiplexing (TDM) technology divides the transmission line into different channels and assigns a time slot for data transfer to every channel associated with each transmitter (Figure 1.20). This resource division scheme is called multiplexing and is associated with layer 1.

T1 lines, specifically, consist of 24 channels and their maximum transmission rate and signaling can be calculated as follows:

Each T1 frame consists of 24 channels of 8 bits each plus 1 bit signaling per frame, and each frame leaves every 125 μs; thus, 8000 frames are generated in 1 s. The calculation is as follows:

(24 channels × 8 bits/channel) = 192 bits/frame + 1 bit/signaling frame = 193 bits/frame * 8000 frames/s = 1,544,000 bps = 1.5 Mbps

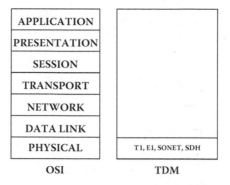

| APPLICATION |
| PRESENTATION |
| SESSION |
| TRANSPORT |
| NETWORK |
| DATA LINK |
| PHYSICAL |

T1, E1, SONET, SDH

OSI   TDM

Figure 1.20 TDM (T1, T3, E1, E3, SONET, SDH) architecture.

Such T1 lines are used in North America while in Europe the lines used are E1. In the latter case each frame consists of 32 channels of 8 bits each and every frame is generated every 125 μs.

For an E1 line the transmission rate with data and signaling is given by the following equation:

(32 channels × 8 bits/channel) = 256 bits/frame * 8000 frames/s = 2,048,000 bps = 2.048 Mbps

In both T and E carriers there are other fairly commercial standards such as T3, with a maximum data and signaling rate of 44.736 Mbps, and E3, with a rate of 34.368 Mbps. There are other T and E specifications that are used in some countries and are complementary to the ones discussed in this book.

There are other technologies based on TDM that can achieve faster speeds, such as Synchronous Optical Networking (SONET) used in North America and Synchronous Digital Hierarchy (SDH) used in the rest of the world. Both technologies, as their name says, perform synchronous transmission between transmitter and receiver.

In SONET, the first carrier layer, OC-1, corresponds to a total transmission rate of 51.84 Mbps, and in SDH STM-1 corresponds to a total transmission rate of 155 Mbps; the same transmission rate in SONET is given by carrier OC-3. The next value in SONET is associated to OC-12, and in SDH STM-4, whose total transfer rate is 622 Mbps. In the case of last-mile, high-speed accesses, rates OC-3 and OC-12 would be used in SONET and STM-1 and STM-4 would be used in SDH. The next transfer rates, OC-48/STM-16, with a rate of 2.4 Gbps, and OC-192/STM-64, with a rate of 9.9 Gbps, belong to the backbone of MAN or WAN carrier networks. Still, there are other standards for SONET and SDH that are not as popular as the ones just mentioned. It is also possible to transmit both SONET

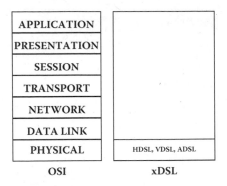

Figure 1.21 xDSL architecture.

and SDH over Wavelength Division Multiplexing (WDM), thus increasing the transmission rate. WDM is discussed in a later section.

Summarizing, the lines T1, T3, E1, E3, SONET, and SDH correspond to layer 1 of the OSI model.

1.3.4.2 xDSL

Digital Subscriber Line (DSL) technologies provide data transmission at rates higher than 1 Mbps, and their main characteristic is that they use the wires of local telephone networks. Examples of DSL technologies include HDSL High bit-rate Digital Subscriber Line (HDSL), Very high bit-rate Digital Subscriber Line (VDSL), Very high bit-rate Digital Subscriber Line 2 (VDSL2), Single-pair High-speed Digital Subscriber Line (SHDSL), and Asymmetric Digital Subscriber Line (ADSL) with the newer versions ADSL2 and ADSL2+ (Figure 1.21).

Table 1.1 is a comparison table of the different DSL technologies.

Table 1.1 Comparison of DSL Technologies

Characteristic	HDSL	VDSL	VDSL2	SHDSL	ADSL	ADSL2	ADSL2+
Speed (Mbps)	2	52 Dn 12 Up	250 100 50 1–4	2.3 SP 4.6 DP	8 Dn 1 Up	12 Dn 1 Up	24 Dn 2 Up
Distance (Km)	4	0.5–1	Out 0.5 1 4–5	3–4	2	2.5	2.5

Note: SP = Single Pair, DP = Double Pair, Dn = Down.

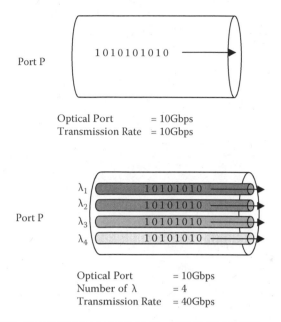

Figure 1.22 (Top) Transmission without DWDM; (Bottom) transmission with DWDM.

As a consequence of the foregoing speeds, DSL technologies are also used as high-speed, last-mile access solutions, but with the convenience of using the same copper lines as telephone lines. We can also say that VDSL and ADSL2+ technologies are appropriate to receive services such as IPTV (Television over IP).

1.3.4.3 WDM (DWDM)

Just like in technologies with TDM, Dense Wavelength Division Multiplexing (DWDM) technology multiplexes the transmission of different wavelengths over a single optic fiber. In other words, with this multiplexing not only are 1s and 0s carried at the fiber level, but the 1s and 0s are transmitted over every wavelength.

Figure 1.22a illustrates optic fiber transmission without DWDM, and Figure 1.22b illustrates the same optic fiber transmission (provided it is compatible) with DWDM.

In this case, devices are currently being developed to support up to 320 λ and could undoubtedly continue increasing; this means that if the transmission speed of the optical port is 10 Gbps, with DWDM it would be possible to transmit approximately 3.2 Tbps. Typical devices found today are 128 or 160 λ.

SONET, SDH, and Ethernet are currently being transmitted over DWDM; for this reason, DWDM is part of layer 1. Figure 1.23 shows DWDM architecture.

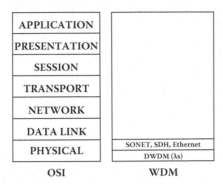

Figure 1.23 DWDM architecture.

1.3.4.4 PPP/HDLC

Point to Point Protocol (PPP) and High Layer Data Link Control (HDLC) are two technologies related to layer 2 of the OSI model that perform the following functions, among others: establishing the communication in a point-to-point network, meaning that such technologies do not perform the switching function; retransmitting packages when there is a loss; recovering in case of package order synchronization failure; performing encapsulation (this function will be explained in detail in a later section); and verifying errors of transmitted bits. The foregoing means that such technologies do not define any layer 1 characteristics and must obviously coexist with some layer 1 technologies. There are many technologies similar to PPP and HDLC that despite having common characteristics are incompatible in function; in other words, if you have a link and connect two routers to this link, one with HDLC configuration and the other with PPP, such routers will not understand each other and, no matter how active the interface is in layer 2, it will not perform its function properly. Similar technologies include LLC and DLC (for LAN networks), LAP-B (2nd layer of the X.25 networks), LAP-F (for Frame Relay networks), and LAP-D (for ISDN networks), among others.

Figure 1.24 shows the architecture of HDLC and PPP. One could, for example, transmit HDLC over HDSL lines or T1 or E1 lines. The same is true for PPP, but in this latter case one has the POS (Packet over SONET) technology that forwards PPP over SONET.

1.3.4.5 Frame Relay

Frame Relay is a technology that operates at layer 2 of the OSI reference model and its main function includes switching, which means that to establish transmission between two edge points in Frame Relay one needs intermediate devices (called switches) that decide where the packages are to be resent. This function is called switching. Like PPP it also performs encapsulation and error control through

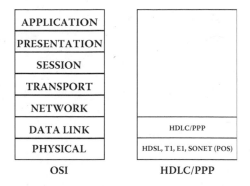

Figure 1.24   PPP/HDLC architecture.

Cyclic Redundancy Code (CRC). Frame Relay does not define layer 1 although at a certain point Integrated Services Digital Network (ISDN) lines had been specified as this layer's solution, but later platforms such as HDSL, line T1, and line E1, among others, were specified.

Figure 1.25 shows the architecture of Frame Relay, which could, for instance, be transmitted over HDSL lines or T1 or E1 lines.

1.3.4.6 ATM

Asynchronous Transfer Mode (ATM) is a network technology that uses the concepts of cell relay and circuit switching and that specifies layer 2 functions of the OSI reference model. What cell relay accomplishes is that the new volume of data to be transferred, called *cell*, has a fixed size, contrary to packages, in which size is variable. ATM cells are 53 bits, of which 48 are data and five are ATM headers. In addition, cell relay technology performs connection-oriented communications and its work scheme is unreliable because the cells are not retransmitted. This

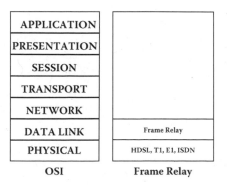

Figure 1.25   Frame Relay architecture.

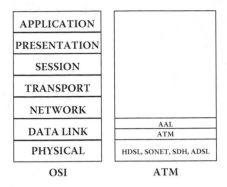

Figure 1.26 ATM architecture.

technology assumes that if the information is lost and needs to be retransmitted, a higher layer will take care of it.

Now all higher layer technologies are specified to work with packages. For this reason it was necessary to specify a sublayer in ATM that will take a package and will fragment it into smaller units in order to construct cells. Similarly, it is necessary to perform the inverse function; that is, when the receiver receives a number of cells, this sublayer must be capable of reconstructing the source package relative to various ATM cells. Another function that must be performed is convergence of services: data, voice, and video convergence was specified in ATM networks, hence it was necessary to establish a function that would identify the different services and integrate them into a single transmission network. All of the functions discussed in this paragraph are performed by a sublayer called *ATM Adaptation Layer* (AAL).

Figure 1.26 shows the architecture for ATM and AAL and, for example, could transmit over HDSL, SONET, SDH, and ADSL, among others.

1.3.4.7 WiMAX

WiMAX technology, whose specification is given by standard IEEE 802.16 and whose solutions span a MAN network, is currently penetrating the markets and there are still uncertainties about its being able to meet the expectations that have risen around it.

Standard IEEE 802.16 in general defines both layer 1 and layer 2 according to the specifications of the OSI reference model. In layer 1 it specifies the type of coding to be used, and here we can mention QPSK, QAM-16, and QAM-64, and defines the time slots through TDM. In layer 2 it specifies three different sublayers: security sublayer, access to the medium sublayer, and convergence of specific services to layer 2 sublayer. In the latter case one can define different types of services and, last, at this layer it defines the frame structure (Figure 1.27).

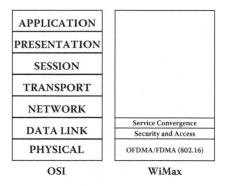

APPLICATION	
PRESENTATION	
SESSION	
TRANSPORT	
NETWORK	
DATA LINK	Service Convergence Security and Access
PHYSICAL	OFDMA/FDMA (802.16)
OSI	WiMax

Figure 1.27 WiMAX architecture.

1.3.4.8 GMPLS

Multiprotocol Label Switching (MPLS) is a data-carrying technology located between layer 2 (link) and layer 3 (network), according to the OSI reference model, that was designed to integrate traffic of the different types of services such as voice, video, and data (Figure 1.28). It was also designed to integrate different types of transportation networks such as circuits and packages and, consequently, to homogenize layer 3 (IP) with layer 2 (Ethernet, WiFi, ATM, Frame Relay, WiMAX, PPP, and HDLC, among others). Through this technology it seems as if IP were connection oriented by means of the creation of the Label Switched Paths (LSPs).

What Generalized Multiprotocol Label Switching (GMPLS) does is expand the control plane concept of MPLS to cover what is defined in MPLS such as SONET/SDH, ATM, Frame Relay, and Ethernet and, in addition to supporting the package commutation technologies, also support time-based multiplexing (TDM) or commutation based on wavelength (λ) or ports or fibers. In this sense, the GMPLS control plane would encompass platforms of both layer 2 and layer 1 (Figure 1.29).

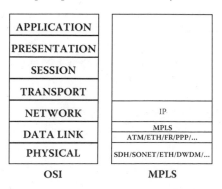

APPLICATION	
PRESENTATION	
SESSION	
TRANSPORT	
NETWORK	IP
DATA LINK	MPLS ATM/ETH/FR/PPP/...
PHYSICAL	SDH/SONET/ETH/DWDM/...
OSI	MPLS

Figure 1.28 MPLS architecture.

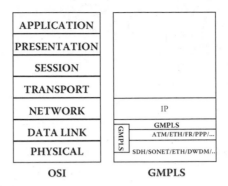

Figure 1.29 GMPLS architecture.

Hence, we can locate GMPLS vertically covering part of layer 2 and part of layer 1, depending upon which platform to which it would be applied.

1.3.5 TCP/IP

The set of TCP/IP protocols allows communication between different machines that execute completely different operating systems. It was developed mainly to solve interoperatibility problems between heterogeneous networks, allowing that hosts need not know the characteristics of intermediate networks.

The following is a description of the four layers of the TCP/IP model.

Network Interface Layer—The network interface layer connects equipment to the local network hardware, connects with the physical medium, and uses a specific protocol to access the medium. This layer performs all the functions of the first two layers in the OSI model.

Internet Layer—This is a service of datagrams without connection. Based on a metrics the Internet layer establishes the routes that the data will take. This layer uses IP addresses to locate equipment on the network. It depends on the routers, which resend the packets over a specific interface depending on the IP address of the destination equipment. It is equivalent to layer 3 of the OSI model.

Transportation Layer—The transportation layer is based on two protocols. User Datagram Protocol (UDP) is a protocol that is not connection oriented: it provides nonreliable datagram services (there is no end-to-end detection or correction of errors), does not retransmit any data that has not been received, and requires little overcharge. For example, this protocol is used for real-time audio and video transmission, where it is not possible to perform retransmissions due to the strict delay requirements of these cases. Transmission Control Protocol (TCP) is a connection-oriented protocol that provides reliable data transmission, guarantees exact and orderly data transfer, retransmits data that

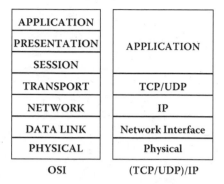

Figure 1.30   TCP/IP architecture.

has not been received, and provides guarantees against data duplication. TCP does the task of layer 4 in the OSI model. All applications working under this protocol require a value called *port*, which identifies an application in an entity, and through which value the connection is made with another application in another entity. TCP supports many of the most popular Internet applications including HTTP, SMTP, FTP, and SSH.

Application Layer—The application layer is similar to the OSI application layer, serving as a communication interface and providing specific application services. There are many protocols on this layer, among them FTP, HTTP, IMAP, IRC, NFS, NNTP, NTP, POP3, SMTP, SNMP, SSH, Telnet, etc.

Figure 1.30 illustrates the TCP/IP model and its layers.

1.4 Network Functions

This section provides an introduction to fundamentals of some of the main functions performed during data network transmission. Among these functions we mention encapsulation, switching, and routing.

1.4.1 Encapsulation

This function consists of introducing data from a higher layer (according to the OSI model) in a lower layer, for example, if an IP package is going to be transferred through a PPP link. Figures 1.24 and 1.30 show that PPP is in layer 2 and IP is in layer 3; therefore, PPP would encapsulate the IP package. Figure 1.31 shows an example of encapsulation.

Figure 1.32 demonstrates the encapsulations from the application layer through to the link layer if we analyze the complete encapsulation process for transmission of a voice package via VoIP in such a PPP channel.

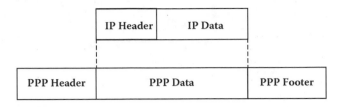

Figure 1.31   Encapsulation.

1.4.2 Switching

The switching function is performed by certain devices when making the decision—given an entry frame with a switching value that can be a label in MPLS, a Data Link Control Identifier (DLCI) in Frame Relay, a Virtual Path Identifier (VPI)/Virtual Channel Identifier (VCI) in ATM, a Medium Access Control (MAC) address in Ethernet, or a λ in Optical Lambda Switching (OLS)/Optical Circuit Switching (OCS)—of which switching value to use, in case the specific technology allows it, and through which physical outlet port such a frame should exit.

To perform switching, technologies such as Frame Relay and ATM, among others, a virtual circuit must first be established that communicates two external switches through a complete network of switches. Such virtual circuits may be created in two ways: Permanent Virtual Circuit (PVC) or Switched Virtual Circuit (SVC). One characteristic of PVCs is that such virtual circuits are always established, whether or not there is data transmission. SVCs, on the other hand, are established while data transmission is taking place. Figure 1.33 shows the concept of virtual circuit, which follows an established path through several switches in the network. In this case it is possible to use several routes, but only one of them has been selected as the transmission route through the virtual circuit.

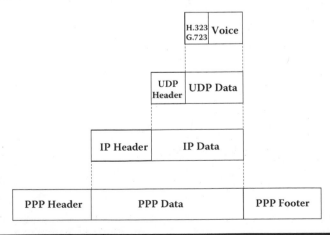

Figure 1.32   Complete encapsulation via VoIP.

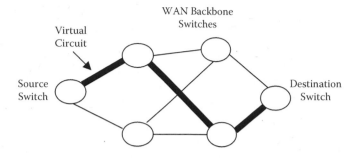

Figure 1.33 Virtual circuit.

In Frame Relay, switching takes place through the DLCI value, which is configured along a virtual circuit. A switching scheme in a Frame Relay network is shown in Figure 1.34. In this example it is possible to transmit information from PC1 to PC2 or from PC2 to PC1 in their respective LAN networks, which are connected through a MAN or WAN network in Frame Relay. To exemplify this case, assume that right now PC1 wants to transmit a package (which can be IP in layer 3) to PC2. PC1 will forward an Ethernet frame placing its own address (PC1) as source MAC address and the MAC address of router R1 as destination MAC address. Based on the destination IP address (this will be explained in more detail in Chapter 3) associated to the PC2 device, router R1 will perform its routing function and decide to forward the package through the last-mile connection. Since the last-mile connection is in Frame Relay, R1 will perform the encapsulation in Frame Relay placing 20 as the DLCI value. Such DLCI values are assigned by the carrier. This frame will follow the logical path marked by the virtual circuit. Next, the frame with DLCI 20 will arrive at the Frame Relay switch of carrier SW1 through the last mile via port P1. Then, SW1 through its switching table will perform the switching function, changing value DLCI 20 for value DLCI 21 (as indicated in SW1's switching table), and forward out the frame via port P2. Following the logical path of the virtual circuit, the frame will come in at SW2 via port P1, where switching will take place again from DLCI 21 to 22, and will be resent out via port P2. Subsequently, SW3 will receive the frame with a value of DLCI 22 and reforward it via port P2 with value DLCI 23 until it finally reaches its destination network through the last mile with router R2. Router R2 will perform its routing function, which will be explained later, to remove the layer 2 Frame Relay header and will replace it with an Ethernet header with the Ethernet destination MAC address of PC2 for delivery thereto.

In ATM, switching takes place through values VPI and VCI, which are configured along a virtual circuit. Figure 1.35 shows switching in an ATM network. In this case it is possible to transmit data from PC1 toward PC2 or from PC2 toward PC1 in their respective LAN networks, which are connected through a MAN or WAN network in ATM. To exemplify this case, assume that right now

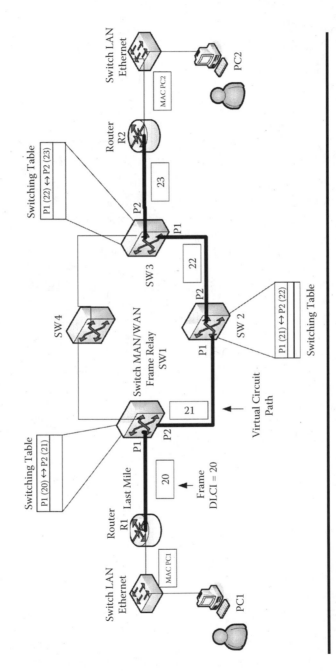

Figure 1.34 Switching in Frame Relay.

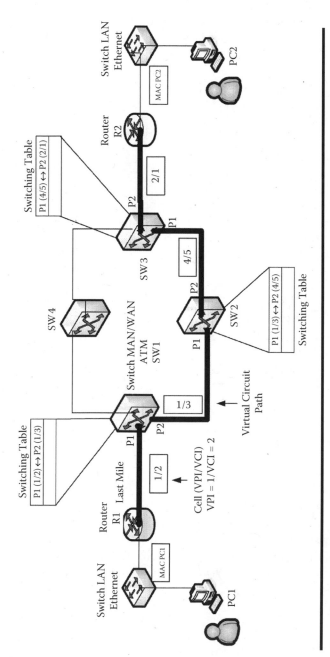

Figure 1.35 Switching in ATM.

PC1 wants to transmit a package (which can be IP in layer 3) to PC2. PC1 will forward a frame placing its own address (PC1) as Source MAC address and the MAC address of router R1 as destination MAC address. Based on the destination IP address (this will be explained in more detail in Chapter 3) associated to device PC2, router R1 will perform its routing function and decide to forward the package through the last-mile connection. Since the last-mile connection is in ATM, R1 will perform the encapsulation in ATM and construct cells assigning VPI/VCI 1/2. Such VPI/VCI values are assigned by the carrier. Said cell will follow the logical path marked by the virtual circuit. Next, the cell with VPI/VCI 1/2 will arrive at the ATM switch of carrier SW1 through the last mile via port P1. Then, SW1 through its switching table will switch the values VPI/VCI 1/2 for values VPI/VCI 1/3 (as indicated in SW1's switching table), and forward out the frame via port P2. Following the logical path of the virtual circuit, the frame will come in at SW2 via port P1, where switching will take place again from VPI/VCI 1/3 for 4/5, and will be resent out via port P2. Subsequently, SW3 will receive the cells with values VPI/VCI 4/5 and reforward via port P2 with values VPI/VCI 2/1 until finally arriving at the destination network through the last mile with router R2. Router R2 will perform its routing function, which will be explained later, to remove the layer 2 ATM header and will replace the Ethernet header cells with the Ethernet destination MAC address of PC2 for delivery thereto.

In MPLS or in GMPLS, as in ATM and Frame Relay, it is first necessary to establish a logical path to transfer data and then perform the switching function. Such logical paths in MPLS/GMPLS are called LSPs (Label Switched Paths). Edge switches that receive an IP package and then retransmit it as an MPLS frame are called LER (Label Edge Router), and switches that are part of the device's network core, that is, those that perform the switching function, are called LSR (Label Switch Router).

In MPLS, switching takes place through the value of LABELs, which are configured along an LSP. Figure 1.36 shows a switching scheme in an MPLS network. In this case, it is possible to transmit data from PC1 to PC2 or from PC2 to PC1 in their respective LAN networks, which are connected through a MAN or WAN network in MPLS. To exemplify this case, assume that right now PC1 wants to transmit a package (which can be IP in layer 3) to PC2. PC1 will forward an Ethernet frame placing its own address (PC1) as source MAC address and the LER1 MAC address as the destination MAC address. Based on the destination IP address (this will be explained in more detail in Chapter 3) associated to device PC2, LER1 will perform its routing function and decide to forward the package through the last-mile connection. Since the last-mile connection is in MPLS, LER1 will perform the encapsulation in MPLS assigning 100 as the LABEL value. Such LABEL values are assigned by the carrier. This frame will follow the logical path marked by the LSP. Next, the frame with LABEL 100 will arrive at the MPLS switch of carrier LSR1 through last mile via port P1. Then, LSR1 through its switching table will perform the switching function, switching LABEL 100 value for a LABEL 200

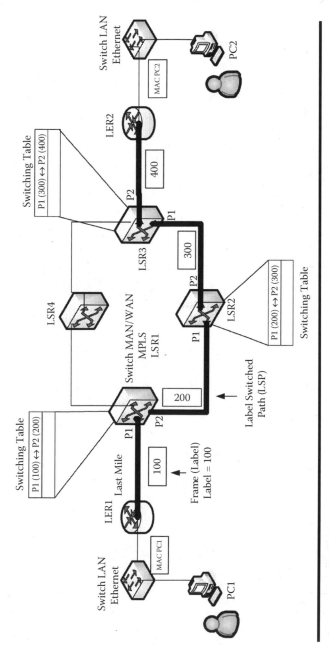

Figure 1.36 Switching in MPLS.

value (as indicated in LSR1's switching table) and forward out the frame via port P2. Following the path of the LSP, the frame will come in at LSR2 via port P1, where switching will take place again from LABEL 200 to 300, and will be resent out via port P2. Subsequently, LSR3 will receive the frame with LABEL 300 and reforward it via port P2 with LABEL 400 value until it finally reaches its destination network through the last mile with LER2. The LER2 will perform its routing function, which will be explained later, to remove the layer 2 MPLS header and will replace the Ethernet header with the Ethernet destination MAC address of PC2 for delivery thereto.

When the switching function takes place in a network with optical switching, such function takes place at layer 1 of the OSI reference model. In GMPLS the switching takes place through the value of the wavelengths (λ), which are configured along an LSP. Figure 1.37 shows a switching scheme in a GMPLS network with OLS. At this point, it is important to clarify that there are different types of optical switching, among them OLS/OCS, Optical Packet Switching (OPS), and Optical Bursa Switching (OBS), but for introductory purposes of the switching concept, only one example of OLS will be provided. A later chapter shows the operation of OPS and OBS. Picking up on the MPLS example, in this case one can transmit data from PC1 to PC2 or from PC2 to PC1 in their respective LAN networks, which are connected through a MAN or WAN network in GMPLS. To exemplify this case, assume that right now PC1 wants to transmit a package (which can be IP in layer 3) to PC2. PC1 will forward an Ethernet frame placing its own address (PC1) as source MAC address and the MAC address of LER1 as destination MAC address. Based on the destination IP address (this will be explained in more detail in the Chapter 3) associated to device PC2, LER1 will perform its routing function and decide to forward the package through the last-mile connection. Since the last-mile connection is in GMPLS, LER1 will transmit the bits through the LAMBDA (λ) λ_1 value. Such λ values are assigned by the carrier. These bits will follow the logical path marked by the LSP. Next, the bits being transmitted, $\lambda \lambda_1$, will arrive at switch GMPLS of the LSR1 through the last mile via port P1. Then LSR1 through its switching table will perform the switching function, switching the value $\lambda \lambda_1$ for value $\lambda \lambda_2$ (as indicated in LSR1's switching table) and forward out the bits via port P2. Following the LSP path, the bits come in at LSR2 via port P1, where switching will again take place from $\lambda \lambda_2$ to λ_3, and will be resent out via port P2. Subsequently, LSR3 will receive the bits with value $\lambda \lambda_3$ and will reforward them via port P2 with value $\lambda \lambda_1$ until finally they reach the destination network through the last mile with LER2. LER2 will perform routing, which will be explained later, to reassemble the bits as a frame and add an Ethernet header with the Ethernet destination MAC address of PC2 for delivery thereto. In this example we have omitted the GMPLS control plane part because we want to show only the switching scheme in this section.

Lastly, in technologies such as Ethernet, the switching technology currently takes place without creating a virtual circuit or LSP. The switching function is

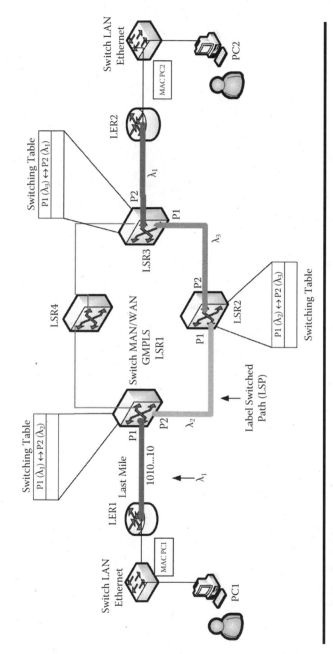

Figure 1.37 Switching in OLS.

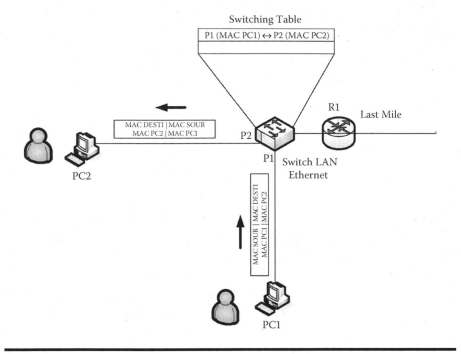

Figure 1.38 Switching in Ethernet.

based on the MAC address of the destination device's network card; in other words, the switching tables of the switches record through which physical port each of the device's MAC addresses is.

Figure 1.38 shows a switching scheme in an Ethernet network. As in all the previous cases, PC1 will transmit a frame to PC2, but in this case PC1 and PC2 are in the same LAN network and even in the same broadcast domain; that is, in the same VLAN (Virtual LAN), in case there are VLANs implemented in the LAN network. In this case, PC1 will include two fields in the frame to be transmitted within the Ethernet header, where it will place MAC PC2 (whose value corresponds to the MAC address of the network card found physically in PC2) as destination MAC address and will place MAC PC1 (whose value corresponds to the MAC address of the network card found physically in PC1) as source MAC address. Next, this frame will arrive at the Ethernet switch via physical port P1 and will seek in the corresponding switching table the PC2 MAC address (MAC PC2). Once found, it will identify the physical port for such MAC address, which in this case would be PC2. Lastly, the switch will reforward the frame with destination PC2 via port P2. This is how switching currently takes place in Ethernet, and we can conclude that to perform this switching there is no need to construct a virtual circuit or LSP.

The foregoing switching scheme is inconvenient for the new challenges and uses being given to this technology such as those for aggregation networks and core by data operators. Standard 802.3ah (from the Institute of Electrical and Electronic Engineering) standardizes client network access to MAN or WAN networks of data operators. Even so, many data operators have implemented Ethernet as an aggregation or core solution for their networks. The problem with this is that the switching function is well defined for the LAN environment, but is not as appropriate for the MAN environment and so much less is for WAN. For this reason, research is taking place worldwide for the construction of LSPs in Ethernet or at least for a switching mechanism similar in technology to MPLS/GMPLS.

1.4.3 Routing

This function takes place when we must transmit data end-to-end through different networks. This means that in this case it is possible to pass through different layer 2 network technologies. Here, it is also necessary to establish a path between the end-to-end networks so that the device in the source network can forward the data to the device in the destination network. Consequently, the routing function takes place in layer 3 of the OSI reference model. IP, the protocol used in the Internet, is a transmission protocol that operates at this layer.

In routing, data can be transmitted to different locations. The first type of transmission is called *unicast*, where the data to be transmitted goes to a single destination device; in *multicast* we want to send the same data to a group of destination devices; in *anycast* we want to send data to any of the destination devices; and, last, in *broadcast* we wish to send the data to all the other devices.

Figure 1.39 illustrates a routing scheme based on IP that shows how it is possible to pass through different layer 2 technologies. In this case, we want to transmit data from a PC1 that is located in a LAN network through router R1 to PC2 that is located on another LAN network through R2 as the destination router. In order to forward such data from router R1 to router R2, it is necessary to establish an established path using a routing protocol such as RIP, OSPF, EIGRP, IS-IS, or BGP, among others. The thick line shows the selected path.

Figure 1.40 shows the same scheme as in Figure 1.39 but with the corresponding routing tables. In this case, we want to transmit a package from PC1, which is located on a source network, to PC2, which is on a different destination network. For the transmission to take place, it is necessary to follow these steps: First, PC1 sees that it is on a different network from the destination device (address IP PC2 30.0.0.1; for more information about IP routing see [COM05]); therefore, through switching, in case there is a switch at the connection between PC1 and router R1, or through physical broadcast, should there be a Hub, the data is forwarded from PC1 to its R1 link port via Ethernet, as shown in Figure 1.38. R1 sees that the IP address of the destination network (30.0.0.0) does not belong to its direct connection and consults its routing table. The R1 routing table shows that to reach network 30.0.0.0 the package must be resent to R3, and R1 it out through physical port P1. Then, R3 also sees that it does not have a direct connection to network

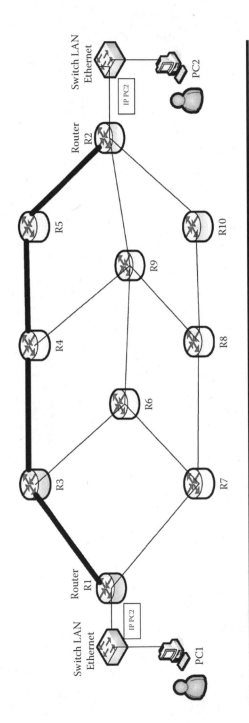

Figure 1.39 Routing with IP.

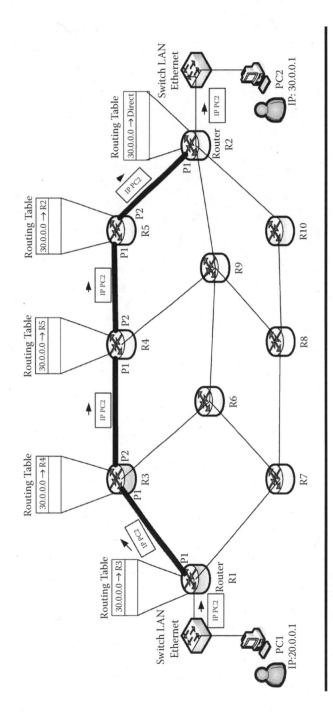

Figure 1.40 Routing table.

30.0.0.0 and therefore consults its routing table to find that it must resend the package to R4 via physical port P2. R4 follows the exact same procedure, and consulting its routing table resends the package to R5 via physical port P2. Next, R5 resends it to R2. Finally, R2 consults its routing table and sees that it has direct connection to network 30.0.0.0; hence, the package will be delivered to PC2 through Ethernet at layer 2, whether through switching, should there be connection switches to the LAN network, or broadcast, if there are hubs in the LAN network.

Figure 1.41 shows an encapsulation, layer 2 switching, and routing integration scheme. With this figure we can begin to see what type of layer 2 technology exists between each pair of routers, that is, the type of link layer. For example, what exists between R1 and R3 is an end-to-end technology that only performs encapsulation and does not perform switching; in this case, PPP is being used with a link protocol. Between R3 and R4 we see a technology that performs switching. Therefore, the thick line means that at layer 2, in this case ATM, it is necessary to establish a virtual circuit through the ATM network for IP packages to travel from R3 to R4. A similar situation occurs between R4 and R5, but here it is necessary to establish an LPS since it is on an MPLS network, and, finally, between R5 and R2 once again we find end-to-end technology, which performs encapsulation but not switching and for this case is HDLC. One must remember that in both LAN networks where the PC1 and the PC2 are located, the layer 2 technology is Ethernet, which performs switching but does not establish a virtual circuit or LSP.

Figure 1.42 illustrates a transmission performing routing and switching jointly at those routers that require it. In this example we can see that there are different layer 2 technologies between every pair of routers. At this point, it is necessary to mention that this type of example does not necessarily represent a real case, but there could be a network shaped like this. First, on the LAN (source and destination) networks we have Ethernet as layer 1 and 2 technologies. Between routers R1 and R3 we have PPP as layer 2 and HDSL could be used as layer 1 technology, for example. Between R3 and R4 we have a complete ATM switching network as layer 2 and SONET/SDH could be layer 1. Between R4 and R5 we also have a complete MPLS switching network as layer 2 and SONET/SDH or Ethernet, for example, could be layer 1. Finally, between R5 and R2 we have an end-to-end HDLC as layer 2 and HDSL could be layer 1.

What follows is a description of the complete transmission procedure between PC1 and PC2 performing the necessary switching and routing functions.

Step 1:   PC1 with address IP 20.0.0.1 wants to transmit an IP data package to device PC2, whose destination IP address is 30.0.0.1. PC1 notices that the network address for PC2 (30.0.0.0) is different from its network address (20.0.0.0) and, therefore, through switching on Ethernet (in the case of hubs this would be by means of broadcast) would forward it to its default gateway, which in this case would be R1.

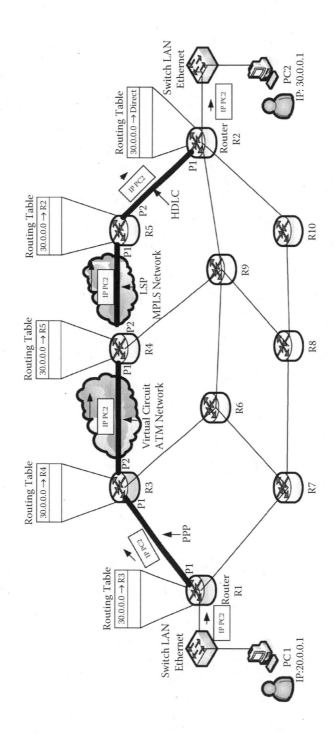

Figure 1.41 Integration of encapsulation, switching, and routing.

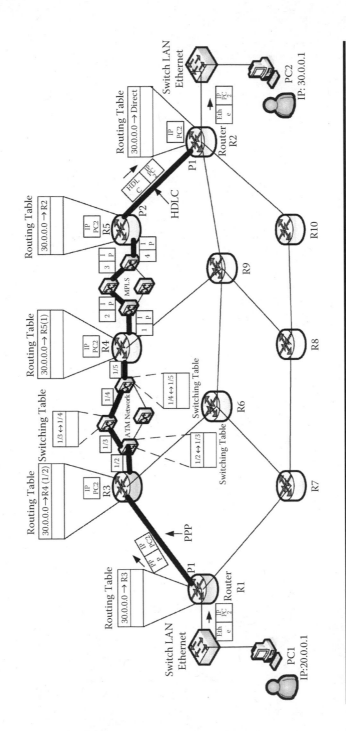

Figure 1.42 Routing and switching.

Step 2:   R1 receives the frame, removes the Ethernet header, and consults the IP address of the destination network (30.0.0.0); R1 consults its routing table and resends the IP package toward router R4 through the ATM network.

Step 3:   R1 fragments the IP package to build 53-byte ATM cells and transmits them through the virtual circuit with a value of VPI/VCI 1/2 as can be observed in R1's routing table.

Step 4:   The switches of the ATM network will perform switching just as in the example shown in the ATM network in Figure 1.35. In this case, the switches will switch the cells with values VPI/VCI 1/2 to 1/3; next, from 1/3 to 1/4 and, finally, from 1/4 to 1/5, as shown in the switching tables in the ATM network.

Step 5:   Router R4 receives the ATM cells corresponding to one IP package and reassembles them.

Step 6:   Router R4 consults the routing table for the destination network address (30.0.0.0) and decides it must forward the package to R5 through an MPLS network and that the switching value in this case is 1.

Step 7:   As in the ATM network, the switches in the MPLS network will switch just like the example shown in the MPLS network in Figure 1.36. In this case, the switches will switch the packages with a LABEL value of 1 to 2, then from 2 to 3 and, finally, from 3 to 4.

Step 8:   Router R5 consults the routing table for the destination network address (30.0.0.0) and decides it must forward the package to R6 via an end-to-end connection through an HDLC encapsulation.

Step 9:   Router R6 receives the MPLS package, removes the MPLS header, checks the destination address (30.0.0.0), and notices that it is connected directly; for this reason, it will add the Ethernet headers to send the package by means of the switching function explained in Figure 1.38 for it to finally arrive at PC2, the destination PC.

1.4.4 Multiplexing

Multiplexing is the process whereby a device called a *multiplexer* receives multiple inputs (whether physical channels or signals) and combines them into one to send them out through a single outlet. The reverse process, when combined channels or signals are received through a single input and are output in their original form, is called *demultiplexing*. The device that performs this process is called a *demultiplexer*. In practice these devices carry out their functions simultaneously.

There are many forms of multiplexing, among which we will mention Time Division Multiplexing (TDM), Frequency Division Multiplexing (FDM), and Code Division Multiplexing (CDM), and in optical communications there is Wavelength Division Multiplexing (WDM). Access methods based on multiplexing have been created for cases where the medium is shared, for example, in wireless networks (radio), and different devices want to transmit using the same frequency. Such access methods include Time Division Multiple Access (TDMA), Carrier

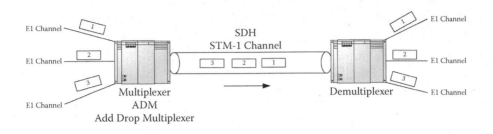

Figure 1.43 TDM.

Sense Multiple Access (CSMA), Frequency Division Multiple Access (FDMA), Orthogonal Frequency Division Multiple Access (OFDMA), and Code Division Multiple Access (CDMA), among others, as there is a wide variety. An example of multiplexing technology applications includes the following: SDH or SONET use TDM, ADSL uses FDM, GSM cellular networks use FDMA and TDMA, 802.11g uses CSMA/CA, and 802.3 uses CSMA/CD.

Figure 1.43 illustrates a multiplexing scheme through TDM. In this case, one can see how the multiplexer receives an input of three E1 channels and multiplexes them through a single channel, STM-1, in SDH. However, the demultiplexer receives channel STM-1 and demultiplexes it through the three E1 channels. The foregoing shows how for each E1 input in the multiplexer it is awarded a time slot for the transmission of its information.

Figure 1.44 shows a multiplexing scheme through FDM. In this case one sees how the multiplexer receives one voice channel, one upstream data channel, and two downstream channels as input. But in this case the information is sent simultaneously, only through different frequencies. For this reason, total transmission capacity is divided among the number of transmitters. The figure shows that voice is transmitted through the F1 frequency range, the upstream data through the F2 range, and the downstream data through the F3 range.

Finally, Figure 1.45 shows a multiplexing scheme through WDM. In this case the OADM receives three physical channels where each one is associated to a λ value and it multiplexes, sending the three λ values through a single physical connection. There are basically two types in WDM: Coarse Wavelength Division Multiplexing (CWDM) and Dense Wavelength Division Multiplexing (DWDM).

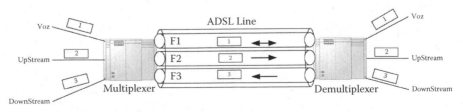

Figure 1.44 FDM.

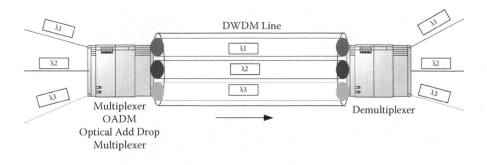

Figure 1.45   WDM.

1.5 Network Equipments

This section explains the most important aspects of the main computer network connection devices.

1.5.1 Hub

This is the most basic connection device in LAN Ethernet networks with which we can connect multiple devices (PCs, servers, printers, etc.) through a shared medium. The hub is a layer 1 device in the OSI reference model. In other words, its basic function is to receive a series of bits (frame) via a port and retransmit it through all other ports. For this reason, said device transmits information in a broadcast form at a physical level. The hub is a device that joins multiple devices together. Therefore, when a device transmits, another device could also be transmitting and the signals would overlap. The effect this produces is called *collision*. For this reason, devices in Ethernet technology support the function known as *Carrier Sense Medium Access with Collision Detection* (CSMA/CD). When hubs are used for network connections this is called *shared Ethernet*, since all have to share the same physical medium to transmit. The fact that hubs are layer 1 implies other consequences such as that the device cannot manage quality of service (QoS) and is therefore unsuitable for transmission of voice, video, or critical data applications. Performance can also be strongly affected when using a shared medium due to the high rates of transmission. In addition, many retransmissions have to take place due to collisions. Another drawback is failure to configure access control lists to establish basic safety levels.

PCs, servers, printers, and other devices are connected to the hub using UTP copper cable with RJ-45 connectors. Several hubs can be connected among them also using a UTP cable or optic fiber.

Figure 1.46 shows an Ethernet hub.

Figure 1.47 shows a transmission scheme through a hub. Figure 1.47a shows when PC1 transmits a package to server S. Because the hub is a layer 1 device, it

Figure 1.46   Hub.

will be unable to know to which specific device the package is addressed and will resend the package through all the other ports, as shown in Figure 1.47b. This operation could result in safety issues in addition to the drawbacks we already mentioned, because if a user on PC1 is transmitting an unencrypted administrator password and another user at a PC other than PC1 and server S installs a Sniffer (software to perform network analysis and to capture packages), he could obtain this password because the hub would also retransmit it to this PC.

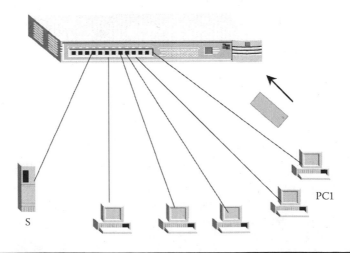

Figure 1.47a   Hub transmission (from PC to Hub).

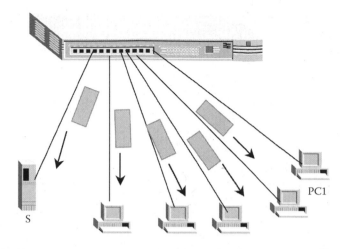

Figure 1.47b   Hub transmission (from Hub to PCs or Server).

1.5.2 Access Point

Access point is the device used to interconnect devices (PCs) wirelessly (see Figure 1.48). In this book we will exclusively discuss the LAN connection access points. As in a hub, the physical medium associated to air is a shared medium and similar problems as those already mentioned could be experienced as well. A protocol called *Carrier Sense Multiple Access with Collision Avoidance* (CSMA/CA) was

Figure 1.48   Access point.

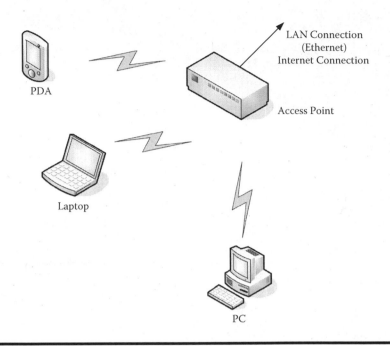

Figure 1.49 Access point connectivity.

created to manage possible collisions when two or more devices want to transmit information. The main purpose of this protocol is not detecting collisions, but preventing them. This type of technology is widely used in the so-called Hot Spots where services such as connection to the Internet are provided. It is normal for an access point to traditionally connect devices in a wireless fashion to the Ethernet wired network. This device is traditionally layer 2. Safety is another issue with shared mediums and in this case, access controls have been implemented at the access points, for example, at the MAC address or the user authentication. Other encryption mechanisms implemented include Wired Equivalent Privacy (WEP), WiFi Protected Access (WPA), and IEEE 802.11i (WPA2). IEEE 802.11 is the standard that specifies operation, with its variety of standards, which include 802.11a (54 Mbps at 5 GHz), 802.11b (11 Mbps at 2.4 GHz), 802.11g (54 Mbps at 2.4 GHz), and the new standard 802.11n (where the idea is to transmit up to 540 Mbps at 2.4 GHz).

Figure 1.49 shows an access point connectivity scheme. In this case, every device, whether a laptop, PC, or PDA, will feature a wireless network card and via WiFi technology will have network access via the access point. One can also see that through the access point one can establish connections to a LAN (Ethernet) network and even to the Internet.

Figure 1.50   Switch.

1.5.3 Switch

In the case of LANs a switch is the most advanced connectivity device in Ethernet by means of which we can connect multiple devices (PCs, servers, printers, etc.) over a medium that operates as dedicated or switched. A switch operates at layer 2 of the OSI model; hence, its basic function is to receive a frame via a port and retransmit it via a specific port where the destination device is located (Figure 1.50).

Switches consist of a switching matrix commonly called *backplane* (Figure 1.51) with a dedicated connection from all input ports to all output ports; that is why all devices connected to the switch are connected in a dedicated fashion. The figure shows a dedicated connection from port IN 1 to port OUT 2; concurrently, there is a dedicated transmission from port IN 3 to port OUT 4. So, if the switch has 1 Gbps input and output ports, it may perform switching of both transmissions simultaneously, contrary to the hub, which is a shared medium and cannot perform such transmissions simultaneously.

When the backplane of a switch has the capacity to switch everything that the input ports send toward the destination ports, it is called a *nonblocking switch*; that is, the switch can resend all packages that arrive through the input ports. Calculating the switching capacity of a switch is simple and is explained next.

Suppose we have a 24-port, 10/100 Mbps switch. This means that the switch can transmit at either 10 Mbps or 100 Mbps. In this case we will analyze the worst case, or 100 Mbps. The worst case for a switching function is when all the IN ports

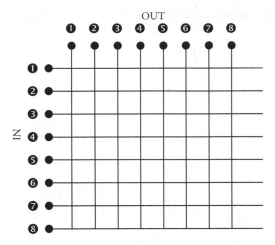

Figure 1.51　Switching matrix.

want to transmit at their maximum rate, which in this example is 100 Mbps. The calculation is as follows: 2 (if transmissions are full duplex) × 24 (ports) × 100 (Mbps) = 4.8 Gbps. If the backplane value that appears on the device's brochure is greater than or equal to 4.8 Gbps, the switch is a nonblocking switch.

Overall, we can say a device's switching capacity is given by

$$\sum_{\text{Port Types}} 2 * \text{Port Quantity} * \text{Port Velocity}$$

Switches can eliminate transmission collisions; still, they will not be eliminated from the network if there is a network where switches and hubs coexist. One of the problems with network design is congestion and the only way to be aware of this problem is by having a clear design. Congestion can happen in a switch when several IN ports want to send information to the same OUT port, as shown in Figure 1.52. Here, IN ports (1, 3, and 4) want to send out their transmission via OUT 2 port. Assuming every port is 1 Gbps, then 3 Gbps would be trying to go out via a 1 Gbps OUT port. To prevent loss of all packages, queues or buffers are implemented. The time packages remain in queue, produce delays in the transmission, and severe congestion results in the loss of packages due to saturation of queues. For this reason a good design and dimensioning of a network is extremely important. This will be shown in other chapters.

Switches may be of different layers depending on the functions they perform in addition to their basic layer 2 switching.

A switch will be layer 2, the lowest layer for a switch, if it performs switching in the LAN case based on the MAC address, which was explained in Section 1.4.2 and is shown in Figure 1.53. In this case the switch will make the decision of resending an

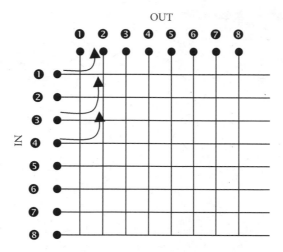

Figure 1.52   Congestion in switches.

Ethernet frame based on its destination MAC address by consulting in the switching table to which OUT port such destination MAC address is associated.

Figure 1.54 shows the switching table of a device, where one can see a record of several MAC addresses.

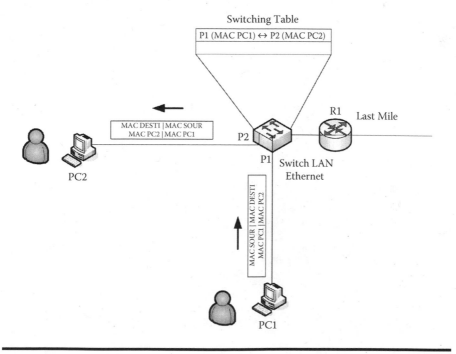

Figure 1.53   Switching layer 2 in Ethernet.

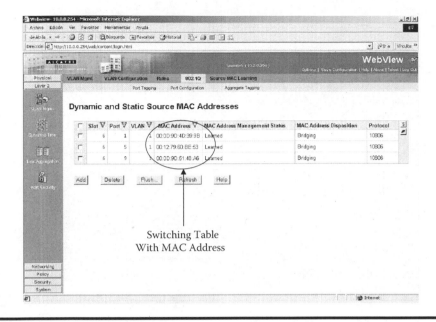

Figure 1.54 Switching layer 2 in Ethernet with equipment.

Figure 1.55 shows by means of the command *ipconfig /all* the values of the IP address and the MAC address (physical address) of a PC.

Figure 1.56 shows how the MAC address of a PC is recorded in the switch's switching table, showing the physical port through which it is found, which in this case is Slot 6 Port 1.

Due to the already mentioned operation, switches will resend the frame through the port only where the destination MAC address is, as shown in Figure 1.57.

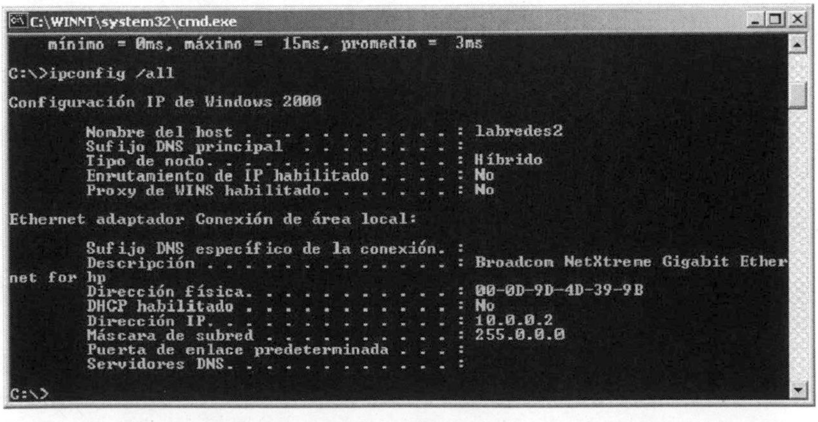

Figure 1.55 Switching layer 2 in Ethernet with equipment.

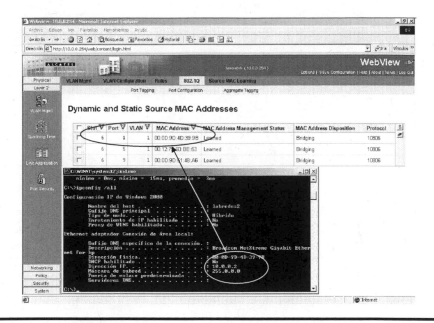

Figure 1.56   Comparison between switching table and MAC address.

A switch that performs a routing function will be classified as layer 3; however, switches perform other layer 3 functions, among them switching layer 3.

First, let us discuss the layer 3 switching function. This function means that the switch has the capacity of switching through the IP address instead of switching through the MAC address, whose value is layer 2. As seen in Figure 1.58, the switch could perform switching based on the MAC address or the IP address and the

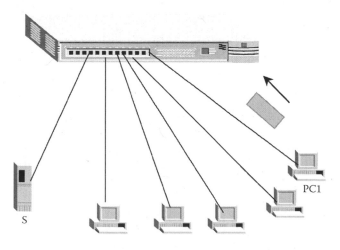

Figure 1.57a   Switch transmission.

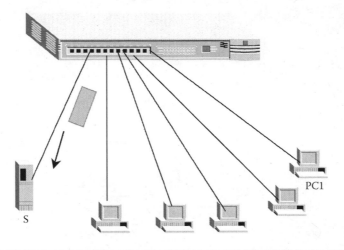

Figure 1.57b   (Continued)

forward time would be exactly the same. The question is, when is layer 3 switching useful? To show this, we will discuss operation of the Address Resolution Protocol (ARP), which consists of finding the MAC address of a device once the destination

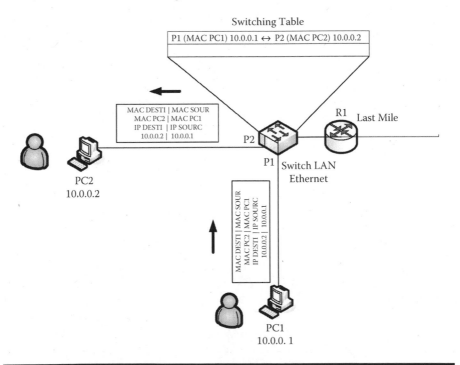

Figure 1.58   Switching layer 3 in Ethernet.

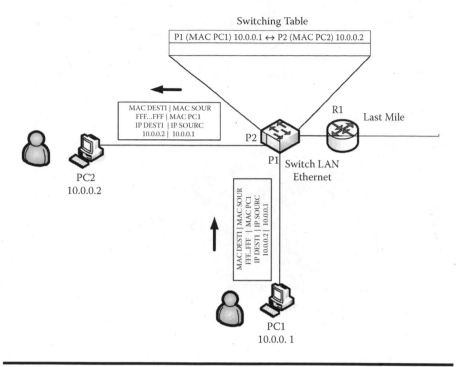

Figure 1.59   Switching layer 3 example.

IP address of such device in the LAN network is given. When the ARP request message is sent, it is sent in such a way that the destination MAC address is broadcast, but the IP address is known; therefore, it is a layer 2 broadcast but a layer 3 unicast transmission. Consequently, if the switch is a layer 2 device, it will send the package through all ports as a broadcast, but if the device is layer 3, it may switch based on the destination IP address. Hence, it will resend the package through a single port that corresponds to where the destination device is. Figure 1.59 shows this ARP case. P1 will transmit a message with a destination MAC address in broadcast (FFF...FFF), but with an IP address (10.0.0.2); therefore, if the switch is layer 3, the switch will resend the package only through where the PC2 (P2) is found.

Figure 1.60 shows a practical case of a switch with the information of the IP address, MAC address, and the physical port through which this device is found recorded in the ARP table.

Now we will explain a second function of layer 3 switches, which is routing. Devices that perform a routing function are normally called *routers* and, although this is entirely true, it does not mean that a switch cannot be a router as well. The question then would be, where would a switch perform a routing function? In the case of a LAN network it would be within what is called a *Virtual LAN* (VLAN) and, in the case of WAN switches, each switch that simultaneously manages

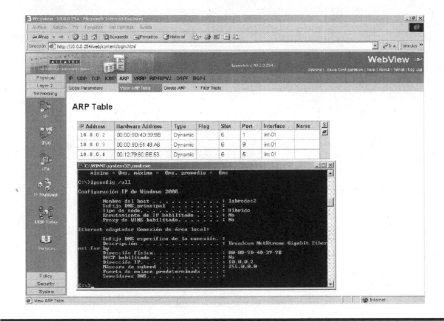

Figure 1.60   ARP table.

technologies such as ATM or MPLS but also manages IP would be considered a router.

In this first part we will continue discussing LAN switches. This book does not intend to provide detailed explanations of the VLAN concept; there are good books that explain it clearly. However, we will provide some fundamentals in order to be able to design LAN networks with VLAN management. The most basic concept that a VLAN can manage, and could be the fundamental reason for its configuration and use in LAN networks, is the division in the broadcast domains. Until now, we have explained that switches separate domains from collision through dedicated communications, but if a switch is sent a message on broadcast, it will resend such message through all the other ports. This means that if only one broadcast domain is available in a LAN network, every time a device sends this type of message it will invade the network completely. If many broadcast messages are generated, this could result in poor network performance due to congestion. It is impossible to eliminate broadcasts from a network because there are many general network applications or functionalities that must be communicated through these message types; for this reason, the only mechanism left to improve performance is isolation or division of such broadcasts through the domains. When VLANs initially came out, they were used for other purposes, which today could be the reason for their implementation in a network. One of these reasons, for example, is for applying safety policies through VLAN; in other words, creating policies that allow or deny transmission of certain applications among VLANs and this way, for example, protect servers.

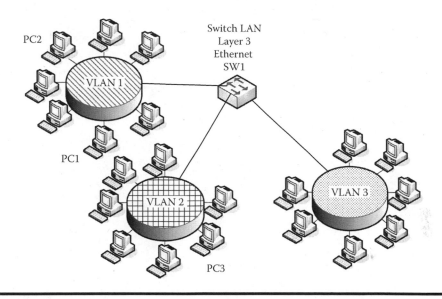

Figure 1.61   Logical scheme of VLANs.

Another reason is for the application of QoS; this way, for example, one can create a VLAN for Voice over IP (VoIP) services or videoconferences and establish a higher priority than the other VLANs, therefore making these services prioritary.

Because different VLAN domains are created through VLANs, this means that two devices that belong to different VLANs cannot see each other during switching; for these devices to see each other there must be routing, for which each VLAN could be specified as an IP network or subnetwork. It is not necessary to have layer 3 switches to create VLANs; they can be configured in layer 2 switches. But, for two devices of two different VLANs to be able to see each other and transmit information, they must be routed, which is precisely the reason for having layer 3 switches in a LAN network, because they perform routing functions that occur through VLANs.

Figure 1.61 shows the logical scheme of VLANs, which demonstrates the concept of *different broadcast domains*, that is, different IP networks or subnetworks. In Figure 1.61 we see that if PC1 wants to communicate with PC2, only the switching function would take place because both PCs are in the same VLAN. But if PC3 wants to communicate with PC1 or with PC2, since PC3 is on a different VLAN, the routing function must take place. In this example the router would be layer 3 switch (SW1), which would be the default gateway for those VLAN devices.

If we analyze what the physical scheme of devices belonging to VLANs would look like, we would have something similar to what is shown in Figure 1.62. Each device that belongs to a VLAN is differentiated by the texture just as in Figure 1.61. We can see that one PC of each of the three VLANs is connected to the main switch, which is layer 3 (the only one that performs routing). The other devices of different

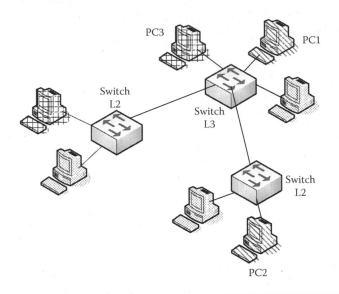

Figure 1.62   Physical scheme of VLANs.

VLANs are connected through layer 2 switches. As this figure shows, the physical location of the device does not matter, provided that connectivity is through Ethernet. Every device will belong to a specific VLAN, and to communicate with another device in the same VLAN, the process can take place only through switching (PC1 with PC2). Also, notwithstanding where two devices are of different VLANs, even if it's in the same switch, as seen in switch L3 (Figure 1.62), for these PCs to communicate a routing function must take place (PC1 with PC3).

A switch will be layer 4 (according to the OSI model) if it understands the information found in the TCP or UDP headers. One of the functions of TCP or UDP is to associate a logical port number to a transmission between client and server applications. That is, a layer 4 switch must understand the value of source port TCP or UDP and the value of destination port TCP or UDP, and this way one can identify what type of application the current package is associated to. Port numbers range from 0 to 65535. Ports 0 through 1023 are reserved by the Internet Assigned Numbers Authority (IANA) for certain services with administrative privileges or TCP or UDP fundamental applications. Lastly, 49152 through 65535 are not established for any application. Table 1.2 shows examples of ports assigned or used in a standard way for applications.

The foregoing means that with layers 1, 2, and 3 of the OSI reference model we can arrive at the destination network and the destination device, but it is with layer 4 that we connect with the specific application at such destination device. A switch with these characteristics will be a device to which one can apply Access Control Lists (ACLs) or QoS characteristics.

Table 1.2   Examples of Ports

Port	Protocol	Application
7	TCP,UDP	ECHO (Ping)
20	TCP	FTP—Data Port (File Transfer Protocol)
21	TCP	FTP—Control Port
22	TCP,UDP	SSH (Secure Shell)
23	TCP,UDP	TELNET
25	TCP,UDP	SMTP (Simple Mail Transfer Protocol)
42	TCP,UDP	WINS (Windows Internet Name Server)
49	TCP,UDP	TACACS (Authentication)
53	TCP,UDP	DNS (Domain Name System)
80	TCP	HTTP (Hyper Text Transfer Protocol)
88	TCP	Kerberos (Authentication)
110	TCP	POP3 (Post Office Protocol version 3)—e-mail
143	TCP,UDP	IMAP4 (Internet Message Access Protocol 4)—e-mail
161	TCP,UDP	SNMP (Simple Network Management Protocol)
179	TCP	BGP (Border Gateway Protocol)
220	TCP,UDP	IMAP (Interactive Mail Access Protocol version 3)
264	TCP,UDP	BGMP (Border Gateway Multicast Protocol)
520	UDP	Routing RIP
646	TCP	LDP (Label Distribution Protocol)
1099	UDP	RMI Registry
1433	TCP,UDP	Microsoft SQL database system
1434	TCP,UDP	Microsoft SQL Monitor
1521	TCP,UDP	Oracle
1812	UDP	RADIUS Authentication Protocol
1863	TCP	Windows Life Messenger

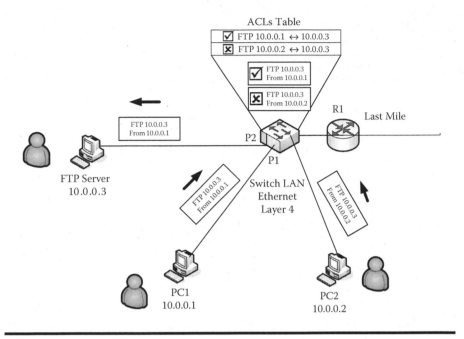

Figure 1.63 Switch layer 4 applying ACLs.

In the case of ACLs, in Figure 1.63 one sees that the ACLs table allows passage of the FTP application from PC1 (10.0.0.1) to the FTP server (10.0.0.3) but denies it for PC2 (10.0.0.2) and, since it is denied for PC2, the package will be rejected by the switch. In this example it is possible to allow, for example, passage of the HTTP application to PC2, for which the switch must understand the information that is in the layer 4 header, that is, TCP or UDP.

In the case of QoS in Figure 1.64, one sees that the QoS table establishes some priority criteria for every type of application. For example, VoIP has priority 6, being the highest in this example; next comes HTTP with priority 3, FTP with priority 1, and the rest of the priority 0 applications (default). In this example there are three PCs that want to transmit information to the Internet through the last mile. PC1 will transmit HTTP, PC2 FTP, and PC3 VoIP. The switch receives the packages and will resend them according to the priority associated to each application as per this example; for this reason, it will first resend the VoIP package, second, the HTTP package, and last, the FTP package. Once again, for the switch to manage the priorities it must understand the value of ports TCP or UDP.

Figure 1.65 shows a configuration associated to layer 4 through ports TCP or UDP to establish conditions, in other words, the way of identifying the flow of information to which QoS or security level access control lists will be applied.

We also find switches that understand beyond layer 4 (TCP or UDP); that is, they understand the application layer. These devices may understand information

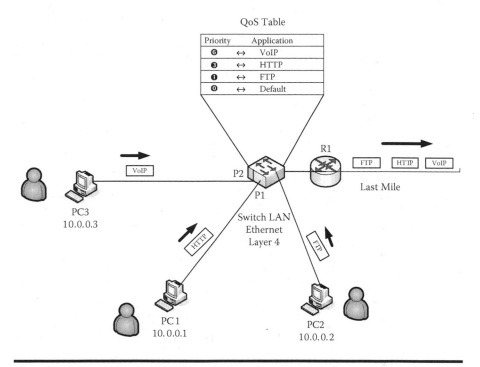

Figure 1.64   Switch layer 4 applying QoS.

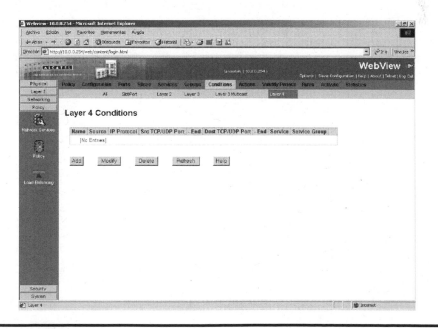

Figure 1.65   Switch layer 4.

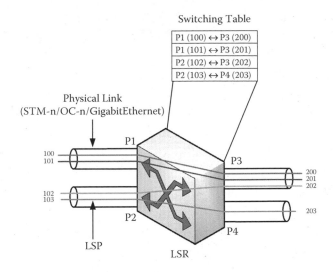

Switching Table

P1 (100) ↔ P3 (200)
P1 (101) ↔ P3 (201)
P2 (102) ↔ P3 (202)
P2 (103) ↔ P4 (203)

Physical Link
(STM-n/OC-n/GigabitEthernet)

Figure 1.66   Switch layer 2 in MPLS.

associated to headers of applications such as Real-Time-Protocol-Transport (RTP) and HTTP parameters, among others. The reason for a switch being in this layer is also associated to ACLs or to QoS.

Finally, regarding LAN switching, we find different types of services for the networks such as ARP proxy, DHCP server, and authentication proxy; one can even find services such as DNS. Such services are added as aggregate services to the basic and advanced operation of switches to somewhat facilitate the number of devices that organizations must take into account for their networking solution.

Most of the functions that we have already mentioned also apply, or some of them with certain differences, for the MAN or WAN switches; an example is establishing the virtual circuit of the LSP to do the switching in WAN switches. Figure 1.66 shows a switching in MPLS scheme where there are four LSPs. For example, when a package with a LABEL 100 arrives at the LSR via physical port P1 (which can be an SDH port, SONET, or GigabitEthernet, for example), the LST will consult its switching table and will resend the package via physical port P3 and will change the LABEL value to 200. The same case would apply to the other LSPs. As we can see, for these types of switches to perform their switching function they must first establish logical paths called LSPs in MPLS. If the LSR is layer 3, then it will also perform routing functions, but in this case such functions would not be associated to VLANs like in Ethernet but to WAN routing. Such switches may perform additional functions that are specified in the manufacturers' brochures for each reference.

Figure 1.67 shows a switching scheme in ATM in which there are four virtual circuits. For example, when a cell with a value of VPI/VCI 3/100 arrives at the switch via physical port P1 (which can be an SDH or SONET port), it will consult

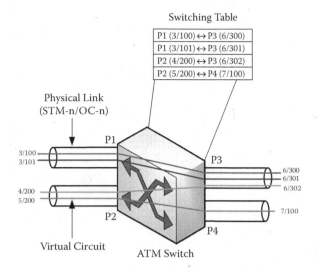

Figure 1.67   Switch layer 2 in ATM.

the switching table and will resend the cells via physical port P3 and the VPI/VCI value will be changed to 6/300. The same would be true for the other virtual circuits. As we can see, for these types of switches to perform their switching function they must first establish logical paths that in ATM are called *virtual circuits*. If the switch is layer 3 it will perform routing functions like in MPLS.

The same would happen in Frame Relay technology, but instead of switching with LABEL or with VPI/VCI, it would switch with DLCI values. Figure 1.68 shows a scheme similar to the ones shown in MPLS and ATM.

1.5.4 Bridge

Bridges are devices that connect to different layer 2 networks. There are also bridges that connect networks with the same layer 2 technology but where segmentation of the network is wanted. In the past, there were bridges that connected an Ethernet network with a Token Ring network, but today switches have all the functions of bridges. Hence, if one wants to purchase a device that performs bridge functions, one will ordinarily find that the device being sold is a switch. It is also important to mention that some functions of bridges are found in routers.

Figure 1.69 shows a typical connectivity associated to a bridge, but which is embedded in a switch.

Figure 1.70 shows a bridge (switch) with Ethernet ports and an ATM/SDH or ATM/SONET port.

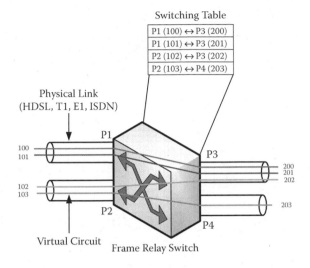

Switching Table

P1 (100) ↔ P3 (200)
P1 (101) ↔ P3 (201)
P2 (102) ↔ P3 (202)
P2 (103) ↔ P4 (203)

Physical Link
(HDSL, T1, E1, ISDN)

Virtual Circuit   Frame Relay Switch

Figure 1.68   Switch layer 2 in Frame Relay.

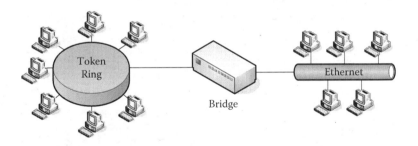

Figure 1.69   Bridge.

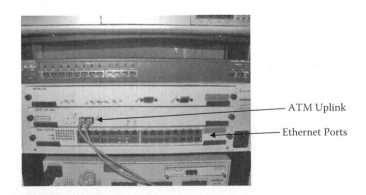

ATM Uplink

Ethernet Ports

Figure 1.70   Bridge Ethernet—ATM.

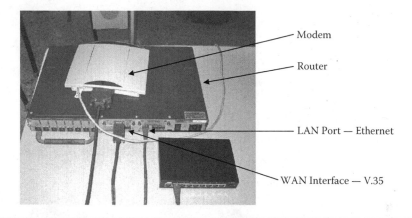

Modem

Router

LAN Port — Ethernet

WAN Interface — V.35

Figure 1.71 Router.

1.5.5 *Router*

In generic terms a router is a device that connects multiple networks by means of the routing function, which was explained in a previous section. Figure 1.71 illustrates a typical router. A layer 3 switch is a router by definition, independent of the layer 2 technology (Ethernet, ATM, Frame Relay, HDLC, PPP, etc.), the layer 1 technology (Ethernet, SONET, SDH, HDSL, etc.), or the interfaces or ports (RJ-45, V.35, etc.) it may exhibit in a specific connectivity case. In the case of the Internet, routers will understand and make their routing decisions based on the destination network's IP address.

One of the consequences of routers connecting to different networks is that for every network they generate a different broadcast domain. In other words, devices in the same network will share broadcasts, but such broadcasts will not be retransmitted to a different network; therefore, for two devices in different networks to see each other, a routing function must be performed.

Figure 1.72 shows the logical scheme of the broadcast domains generated by routers. According to this example, if PC1 generates a broadcast message, such message will reach all PCs located in Broadcast 1 Domain (for example, to PC1 and router R1 via port P1), but it will not reach PC3. The foregoing operation means that PC1 and PC2 can transmit information to each other via switching because they are in the same broadcast domain, but if PC1 or PC2 wants to transmit information to PC3, it must do it through routing in R1.

In addition to the regular functions of a layer 3 device, many manufacturers have added to routers other functions such as firewall, VPN, QoS management, and DHCP servers, among many others.

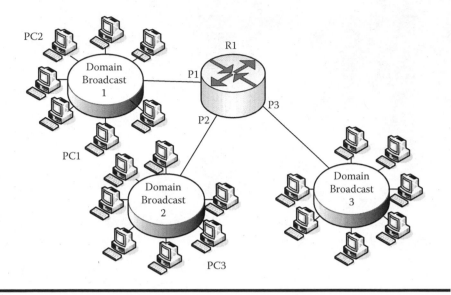

Figure 1.72   Broadcast domain.

1.5.6 Multiplexer

Multiplexers are devices that receive multiple input ports and combine them using a mechanism explained in Section 1.4.4 to output them through a single output port.

In SDH and SONET technology multiplexers are called *Add Drop Multiplexers* (ADMs), in ADSL they are called *Digital Subscriber Line Access Multiplexers* (DSLAMs), and in DWDM they are called *Optical Add Drop Multiplexers* (OADMs).

Figure 1.73 shows an OADM.

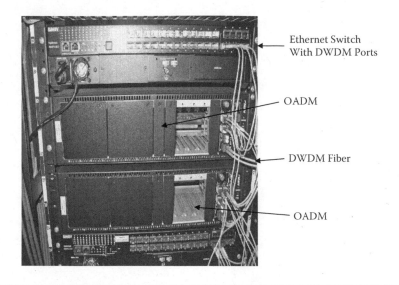

Figure 1.73   OADM.

Chapter 2

LAN Networks Design

This chapter introduces designs and configurations of LAN networks, which are traditionally associated with business networks, college campus networks, and hot spots, among others. Ethernet and WiFi technologies are used to provide LAN solutions. This chapter also discusses VLAN design and the connection scheme of LAN networks to MAN or WAN networks.

2.1 Ethernet Solution

In this section we introduce PC connectivity to edge equipment, connectivity of the different edge equipment to provide solutions due to the number of points needed at that edge, and connectivity toward the network core.

We would like to provide some information about the evolution of Ethernet through time. Standard 802.3 takes off from the Ethernet network standard. In 1980, Digital Equipment Corp. (DEC), Intel, and Xerox (the group that gave rise to the term DIX) created the standard for a 10 Mbps network over coaxial cable using the technique to access medium CSMA/CD and called it Ethernet version 1.0. This technique was later improved in 1985 with Ethernet version 2.0. Project 802 of the IEEE adopted Ethernet version 2.0 as the starting point for its standard 802.3 CSMA/CD. Subsequently, new requirements were needed for the new applications such as multimedia. It is for this reason that standard 802.3u was created and approved in June 1995. The goal of this committee was to maintain the traditional Ethernet transmission protocol, CSMA/CD; support popular cabling schemes (10BaseT); and ensure that fast Ethernet technology would not require changes in protocols of upper layers or in the software that runs in the LAN workstations (i.e., SNMP, MIBs), and would operate at 100 Mbps; use the same type of frames, size, and format; maintain the CSMA/CD; support Ethernet and Token Ring applications; have very simple and cheap design and configuration and strong support for multimedia; and require a new hub and switch adapting card. Fast Ethernet may run through the same variety of media as 10BaseT (UTP, STP, and fiber), but it does not support coaxial cable. It defines three types of medium: 100 Base-TX; category 5, 2 pair UTP and STP cable; and 100 Base-T4. It allows category 3, 4, and 5; 100Base-FX; and optical fiber cable. Subsequently, in 1998, the IEEE published standard 802.3z, in which 1 Gbps is transmitted in optical fiber and, in 1999, standard 802.3ab was approved for transmission of 1 Gbps in copper. This new set of 1 Gbps standards has the following advantages: increased bandwidth and elimination of bottlenecks; wide display capacities using Gigabit over CAT-5 copper and 6, 6e, and 7 cable; full duplex capacity; quality of service (QoS) (audio and video); and low cost of acquisition and ownership. Gigabit Ethernet technology maintains the same specifications as its predecessors: same Ethernet package format, full duplex and half duplex, CSMA/CD protocol, flow control, and management objects as they are defined in standard IEEE 802.3. Gigabit Ethernet can run through optical fiber, but it was established for twisted pairs. It defines six types of media: 1000BaseSX, short-wave optical fiber; 1000BaseLX, long-wave optical fiber; 1000BaseCX, 2 pair STP; 1000BaseT, 4 pair UTP; 1000BaseLH and 1000BaseZX, reaching a distance of up to 100 Km, and IEEE through standard 802.3ae specified transmission at 10 Gbps. These are the specifications of this technology: 10GBase-S, 10GBase-L, 10GBase-E, and 10GBase-LX4, reaching up to approximately 50 Km.

All of these implementations in LAN networks required structured cabling to establish orderly connectivity between the stations and the edge switches and the edge switches and the core. Although the purpose of this book is not to design and specify structured cabling, we will next explain some of the fundamentals of structured cabling.

A structured cabling system (SCS) is basically an information network within a building or group of buildings whose transmission medium is a combination of unshielded twisted pair, UTP, and traditional optical fiber. It is also known by other names, among them intelligent cabling and signal distribution system. This cabling system integrates connection of voice, data, and image equipments, that is, any device whose function is to provide information services through an electrical or luminic signals conductor. In a structured cabling system it is possible to connect switchboards, telephone extensions, computer terminals, video devices, closed television circuits, computer networks, faxes, and so on.

The fundamental purpose of a structured cabling system is to provide users an efficient, inexpensive way of organizing and integrating their information networks, independently of the equipment brands composing such networks, providing as well the possibility of modifying and updating the information network without having to alter the cabling.

Equipment or devices that compose structured cabling include the following:

Patch panel—the central collector of structured cabling

Rack—the device in which hubs, patch panels, switches, etc., are grouped or located

Patch cord—the cable that goes from the terminal to the work area or from the patch panel to the hub or switch

Cross connect—a group of connection points installed on a wall or rack used as mechanical terminations to manage cabling in a building

Transition point—the point in horizontal cabling where a flat cable connects to a round cable

A structured cabling system consists of the following subsystems: building inlet, equipment room, vertical cabling (backbone), telecommunications closet, horizontal cabling, telecommunications outlets, work area, and campus.

Building Entrance. The building entrance provides a point where the external cabling joins the building internal backbone. Standard EIA/TIA 569 defines the physical requirements for this interface. The building inlet consists of a telecommunications services entrance to the building, which includes the entry point through the building wall and continuing through to the entrance room or area. The building entrance should contain the backbone path that interconnects with other campus buildings. In case of communication by means of an antenna, the antenna would also belong in the building entrance.

Equipment Room. The equipment room is a central location for the telecommunications equipment (e.g., PBX, computers, switch) that serve the building's inhabitants. Only equipment that are directly related to the telecommunications system and its support system should be kept in this room. The standard that regulates this subsystem is standard EIA/TIA 569. When selecting the room, avoid locations that are restricted by building components such as elevators, stairs, etc., and it should have accessibility for large equipment. Access to this room should be restricted to authorized personnel only. The floor resistance should be able to sustain the distributed load and the concentrated load of the installed devices. Distributed load should exceed 12.0 kps (250 lbf/ft^2) and concentrated load should exceed 4.4 kN (1000 lbf) over the area of greater concentration of devices. The equipment room should not be located below water level unless measures have been taken to prevent infiltration. The room should have a drain in the event water becomes a problem. The room should also have direct access to the heating, ventilating, and air-conditioning (HVAC) system and should be located away from electromagnetic interference sources, at a distance that reduces interference to 3.0 V/m through the frequency spectrum. Be very careful of electrical transformers, motors, generators, X-ray equipment, and transmission radios or radars. It is desirable to locate the room close to the main backbone. The size of the room

Table 2.1

Number of Workstations	Area in m²
Up to 100	14
From 101 to 400	37
From 401 to 800	74
From 801 to 1,200	111

should be enough to accommodate equipment requirements. To determine the size of the room, take into account current requirements as well as future projects. When size specifications of equipment are not known, the following considerations must be taken into account:

Voice and Data Guide. Practice consists of providing 0.07 m² space in the room for every 10 m² of work area. The equipment room must be designed for a minimum 14 m². Based on the number of workstations, room size should be according to Table 2.1.

Guide for Other Equipment. Environmental control equipment such as energy distributors, air conditioners, and UPA (unit power amplifier) up to 100 kVA must be installed in the equipment room. UPS (uninterruptible power supply) larger than 100 kVA must be located in separate rooms. Minimum height of an equipment room should be 2.44 m (8 feet) without obstructions. The room should be protected from contamination and pollution that could affect operation and the material of the installed equipment. When existing contamination exceeds the level established in Table 2.2, vapor barriers or filters must be installed in the room.

The equipment room must be connected to the backbone route. In case smoke detectors are required, they must be within the original box to prevent

Table 2.2

Contaminant	Concentration
Chlorine	0.01 ppm
Hydrogen Sulfate	0.05 ppm
Nitrogen Oxide	0.01 ppm
Sulphur Dioxide	0.3 ppm
Dust	100 ug/m³/24 h
Hydrocarbon	4 ug/m³/24 h

accidental activation. A drain must be installed under fire detectors to prevent flooding of the room. HVAC equipment must be provided to work 24 hours per day and 365 days a year. If the building system does not guarantee continuous operation, a standalone unit must be installed for the equipment room. Temperature must be regulated in a range between 18°C to 24°C with humidity from 30% to 55%. Humidifiers and dehumidifiers may be required depending on the location's environmental conditions. Room temperature and humidity must be measured at a distance of 1.5 m above ground level and after the equipment is in operation. If backup batteries are used, appropriate ventilation equipment must be installed.

Interior Finishes. The floor, walls, and ceiling must be sealed to reduce dust. Finishes must be in bright colors to increase room lighting.

Flooring material must have antistatic properties.

Lighting. Lighting must have a minimum 540 lx, measured 1 m above ground in a place free of equipment. Lighting must be controlled by one or more switches located near the room's entrance.

Power. A separate circuit must be installed to supply power to the equipment room and this circuit must end at its own electric board. Electric power entering the room is not specified as it depends on the number of pieces of equipment installed.

Door. The door must have a lock and must measure, at a minimum, 910 mm wide and 2.00 mm high. If very large equipment is estimated, then double doors measuring 1.820 mm wide and 2.280 mm high, must be installed.

Grounding. A 1½ conduit must be installed from the equipment room to the building's grounding electrode.

Fire Extinguishers. Portable fire extinguishers must be provided and maintained periodically. They must be installed as close as possible to the door.

Backbone. The backbone provides connection between the telecommunications cabinet, the equipment room, and the building inlet. The backbone consists of the backbone cable, the intermediate and principal cross-connect, the mechanical terminals, and the patch cords.

The Rack. The equipment room and assigned points may be located in different buildings; the backbone includes the transportation medium between different buildings.

The backbone must support all the equipment found in the rack and frequently all the printers, terminals, and file servers for one floor of a building. If more clients or servers are added to a floor, they compete for the available bandwidth in the vertical cabling. However, there is an advantage, which is the small number of vertical channels in a building, which allows the use of more expensive equipment to provide greater bandwidth. This is the area where optical fiber has become the most appropriate medium.

Selecting the Transmission Medium. A backbone for vertical cabling can be built with any of the previously mentioned standards, but the following factors must be taken into account: flexibility regarding the services supported, required backbone shelf life, location size, and user population.

Other Design Considerations. No more than two hierarchical layers of cross-connects can be installed; bridges cannot be used; the length of the patchcord of the intermediate and principal cross-connect cannot be longer than 20 m; grounding must comply with the requirements defined in standard EIA/TIA 607.

Telecommunications Closet. The telecommunications closet is the area inside the building where telecommunications equipment of the cabling system is kept. This includes mechanical terminals and cross-connects for the vertical and horizontal cabling. The norm that standardized this subsystem is EIA/TIA 569. The telecommunications closet on each floor is a transition point between the vertical and horizontal cabling route and must be located as close as possible to the center of the area that it is going to serve.

The telecommunications closet cannot be shared with electrical installations other than those used for telecommunications. There must be, at a minimum, one closet per floor. Additional closets must be installed when floor area is greater than 1000 m^2 and the horizontal distribution distance is larger than 90 m.

Multiple closets on one floor must be interconnected at a minimum by one conduit.

Based on one workstation every 10 m^2, the size of the telecommunications closet must be according to Table 2.3.

The telecommunications closet must be located on floors designed with a load capacity of 2.4 kPa (50 lbf/ft^2). Lighting must be at least 540 lx, measured 1 m above ground. No suspended ceilings should be installed. Door dimensions must be at a minimum 910 millimeters wide and 2000 millimeters high, with a lock. Walls, ceiling, and floor must be treated to eliminate dust. Finishes must be bright to increase room lighting. A minimum of two 15 A and 110 V AC dedicated duplex electrical outlets, each in separate circuits, must be installed to provide power to the equipment.

Table 2.3

Area in m^2	Size of Closet
1000	3000 × 3400 mm
800	3000 × 2800 mm
500	3000 × 2200 mm

It must have access to building grounding. The telecommunications closet must be located in an accessible area in every floor, such as, for example, a common hallway, and access must be restricted to authorized personnel only. The closet must have fire protection and smoke detectors must be installed in their protective cases to prevent accidental activation.

Horizontal Cabling. Horizontal cabling is the portion of the telecommunications cabling system that extends from the telecommunications closet (or rack, which contains the hubs, patch panels, etc.) to the final users' workstations. It consists of (1) horizontal cabling, (2) telecommunications outlet, (3) cable terminators, and (4) cross-connections.

The term horizontal is used because, typically, the cabling system is installed horizontally through the building floor or ceiling. The services that can be transmitted through structured cabling include (1) voice transmission service, (2) data transfer, (3) flexibility relocating devices, and (4) connection to Local Area Network (LAN).

In addition, to satisfy current telecommunications requirements, structured cabling must enable daily maintenance, relocation, and support for the installation of new equipment and services.

Telecommunication Outlets. Every work area must have at least two outlet ports: one for voice and one for data. Different kinds of configuration exist in accordance with certain standards, as follows:

Universal Services Ordering Codes (USOC). USOC are a series of standards for RJ connectors developed by Bell Systems, for connection to public networks.

Configurations of TIA/EIA 568-A. Configuration for LAN networks. Local area network standards designed to operate over UTP designated assignment of pin/par in modular connectors for signal transmission. While TIA/EIA (T568A and T568B) support all these designations, there are certain cases where the user decides to cable only the number or pairs required to support the application.

Work Area. The work area extends from the telecommunications outlet to the workstation. The work area cable is designed for relatively easy interconnection, so that moving, changing, and adding it can be simple. Components in the work area include (1) station devices (computers, phones, data terminals, etc.); (2) patch cables (modular cables, PC adapting cables, fiber jumpers, etc.); and (3) adapters (balloons), which must be external to the telecommunications outlet.

The work area cable may vary depending on the application. A cable with identical connectors at both ends is commonly used. When adaptations are necessary in the work area, they must be external to the outlet. Some of the most common adaptations are as follows: (1) a special cable or adapter is required when the equipment connector is different from the outlet connector; (2) a "Y" adapter is required when two services run in the same multipar cable; (3) a passive adapter may be used when the type of cable in the horizontal cabling is different from the type of cable required by the equipment; and (4) an active adapter may be required when pieces of equipment that use different signal schemes are connected.

In certain cases, pairs may have to be transposed to achieve compatibility.

Certain telecommunications equipment requires resistances in the work area. This can be accomplished by placing the resistance outside the outlet.

The recommendation is to install one outlet every 10m^2.

Finally, for more information on structured cabling specifications, please refer to the following standards. These are standards that were developed to standardize structured cabling:

TIA/EIA 568—Standards for cabling in commercial buildings

TIA/EIA 569—Commercial building standard for telecommunications pathways and spaces

TIA/EIA TSB-36—Specifications for UTP cable

TIA/EIA TSB-40—Specifications for the connection hardware to UTP cables

TIA/EIA TSB-53—Specifications for STP cable

TIA/EIA 607—Standards for grounding in telecommunications equipment

TIA/EIA 570—Residential telecommunications cabling standards

TIA/EIA 606—Administration standard for the telecommunications infrastructure of commercial buildings

IEEE 802.3 and Ethernet/Fast Ethernet LAN standard

IEEE 802.5 and standard Token Ring

ANSI X3T9.5 and FDDI-TP PMD standard

IEEE 802.12 and 100VG-AnyLAN

ATM and ATM forum

A complementary design in new buildings is intelligent buildings, of which LAN networks are part. The following section talks about the purpose of intelligent buildings.

The first concept about intelligent buildings was born in the 1970s, when the first buildings with electronic systems, called *HVAC systems*, were developed. This consisted of a computer with connected sensors that detected changing environmental conditions. The intelligent concepts were first heard in the United States around 1981. In the 1980s, security systems, lighting systems, etc., began to become automated. The first generation of intelligent buildings in London was the Lloyds building, which was designed by Richard Roger Partnership and consisted of individual systems. In the second generation, control was performed by a central computer. For example, a fire alarm system linked with the security system could automatically unlock the doors to provide expedited and easy evacuation. An intelligent building is one that provides a productive and efficient environment by means of an appropriate combination and optimization of its four basic elements: structure, systems, services, and administration.

The intention is to increase technical performance, investment savings, and savings in operation costs, in addition to offering occupants maximum well-being, security, and flexibility. *Structure* refers to the building's physical and structural variables. For example, the distance of the concrete suspended ceiling must be such that access to the cable trays passing through it is fast and simple, in order to avoid future complications in cases of maintenance.

Services refers to the facilities that building occupants have for their comfort: parking, cafeteria, security, etc.

Finally, *systems* are the devices and controls necessary for the services and administration to function correctly. The systems allow the administration to keep appropriate control over power consumption by air conditioning and lighting. The administration can maintain better stock of spares by programming preventative maintenance routines, thanks to the automated control of devices. The building provides enhanced service for personal security and security of goods with subsystems such as access control, closed-circuit TV, and fire detection and control. Although all subsystems in a building normally operate autonomously—it should be this way—centralized control dramatically increases efficiency of these systems. For example, access control together with lighting or air-conditioning control represents the perfect energy savings combination; similarly, the security system linked with the elevators increases security of goods.

The following is a list of an intelligent building's systems and considerations about their function:

Security System. A security system must provide the following services: protection of interior and exterior doors, objects, and specific areas; manual or programmed restricted access; remote supervision and control of security points; remote opening and closing of doors; programming and recording vault, safe, etc., opening and closing; and detection systems due to changes in environmental patterns.

Fire Detection and Alarm. A fire detection and alarm system must provide the following services: continuous supervision and monitoring of all detection devices, alarms, and/or circuit damages; correct management of the main and auxiliary water supply system during emergencies; control of elevator operation; prerecorded messages with instructions; correct management of air conditioners to prevent circulation of toxic gases and asphyxiating fumes; generation of audiovisual alarms; automatic communication with firefighters and other authorities; control of door and stairway pressurization systems; audio communication with building operators; monitoring of water flow in the networks; monitoring of tank water levels; and automatic control of valves and gates.

Control of Access to Restricted Areas. A control of access to restricted areas system must provide the following services: control via magnetic cards, control via proximity cards, recording of all authorized and unauthorized transactions, entry options by hours or days, visitor cards, re-entry control, and additional keyboard for high security points.

Energy Control. An energy control system must provide the following services: control air-conditioning equipment and the interior and exterior lighting system; monitor the power substation, power generator, and consumption.

Communications. At a communications platform level, buildings today must incorporate structured cabling systems that combine copper and optical fiber according to their needs. Development of recent forms of communication such as videoconference, high-speed telefax, multimedia service, interconnection of computers with different characteristics, and local area networks require the use of an advanced structured cabling system capable of supporting all the needs of present day and future communications.

A structured cabling system enables modification of the use of different outlets, in other words, changing the application of a telephone outlet for a data or television outlet, without incurring in physical works. Overall, the structured cabling system allows the building complete interconnection flexibility, which permits multiple uses of space without modifying piping.

A structured cabling system constitutes the distribution network in an intelligent building, integrating into one cabling system a wide range of services and systems such as voice services (analog–digital), data and LAN services, security systems, fire alarm control, air-conditioning control, energy control systems, lighting control systems, expert systems, office automation, maintenance systems, and cabling management and control.

The main purpose when designing an intelligent building is to attain integration of three aspects: automation, communications, and security. Integration of the three systems provides the following advantages: centralized control, resource optimization, process automation, increased productivity, high reliability, and improved living and working conditions.

If each of the systems operates autonomously and integration is accomplished, the following characteristics will be attained: access control, alarm management (electronic surveillance), fire control, telesurveillance (video), and technical management—control (lights, energy, air-conditioning, heating, etc.) of machine flows, metal detection and luggage control, structured cabling system, PABX, tariffication, videoconference, electronic mail, and, overall, voice, data, and video services.

Having explained certain characteristics and considerations about structured cabling and intelligent buildings, we will now explain the connectivity scheme and edge design in a LAN network via Ethernet.

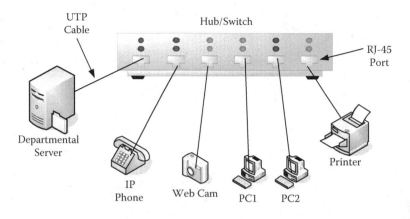

Figure 2.1   Edge connectivity.

2.1.1 Edge Connectivity

Connectivity of devices through Ethernet, in other words, without using a wireless medium, takes place as follows. PCs, IP telephones, Web cameras, network printers, and other devices whose connectivity is through the data network connect via copper pairs (traditionally UTP). Figure 2.1 shows connectivity of equipment toward a hub or edge switch.

Most network connections are actually PC connections more than any other equipment; for this reason, the size of the network is normally calculated through this equipment. Other devices such as IP telephones provide the dimensions for quality of service (QoS) characteristics that we will discuss later.

To provide a solution to an edge, it will probably be necessary to use more than one hub or switch due to the large number of devices that will be connected. In order to be able to supply the number of devices to be connected, perhaps it will be necessary to connect to several hubs or switches or a combination of hubs with switches between them. This connection can be done through ports with larger transmission capacity than are usually used as uplink connection toward the core, or another way, through special ports called *stacks*. Figure 2.2 shows a switch connection scheme via hubs or switches through an uplink port. This uplink port may be via copper pairs or optical fiber, and this factor depends on distance and possible external factors that could affect the physical environment. As you can see, the two edge switches would be covering the number of ports necessary to connect all the devices located in that area of that edge.

In the case of connectivity via a stack port, a special port traditionally located in the back side of hubs or switches is used, and in a logical way gives the appearance of a single device with a greater number of ports. However, the foregoing function can also be performed by new devices when they are connected through the uplink via a management system. Figure 2.3 shows a connectivity scheme through stack

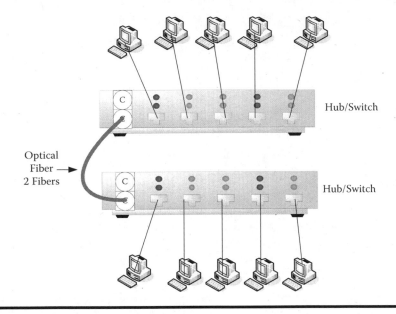

Figure 2.2   Switch connections by uplink.

ports. Most commonly, hubs or switches are manufactured with 24 or 48 ports, although certain brands handle larger or smaller port numbers. That is, if we have an edge that requires 60 ports, one would have to buy three 24-port switches or one 48-port switch and one 24-port switch. Lately, manufacturers have opted more for the connectivity between switches through the uplink port scheme, but this does

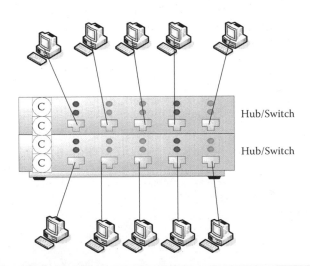

Figure 2.3   Switch connections by stack ports.

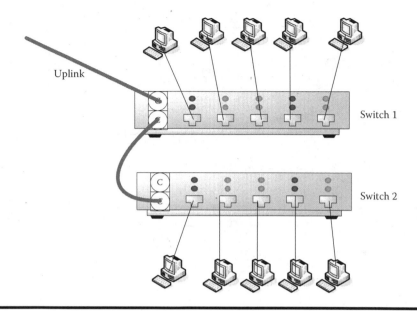

Figure 2.4   Edge-core connection.

not rule out that there are still a large number of switches that connect through stack ports.

Connectivity specifications between computing devices such as PCs and hubs or switches are shown through the concepts of structured cabling, which specify cabling type (in this case horizontal), patch cords, patch panels, and racks, among others—topics that are outside the scope of this book.

To conclude, we can say that the purpose of edge is to meet the needs of the number of pieces of equipment that will be connecting to the computer network, to establish the number of hubs or switches required to meet such demand, and to establish how they will connect among each other at that edge.

The other point to specify is the way connectivity occurs at the core and how many links will be used for such connection. As to the how, this depends, again, on the distance and possibly other factors such as the influence of external factors that could affect transmission over copper. It is very common to use connections in optical fiber from the edge to the core.

Figure 2.4 illustrates an edge-to-core connectivity scheme. In this case, one of the optical fiber ports on switch 1 is used to connect to switch 2 and the other optical fiber port is used to connect to the core. If we use a stack port scheme to connect the switches (switch 1 and switch 2), connection to the core would be exactly the same, that is, through the optical fiber of the uplink port.

Figure 2.5 shows a picture of a real connection case: from a layer 3 edge switch to a core switch.

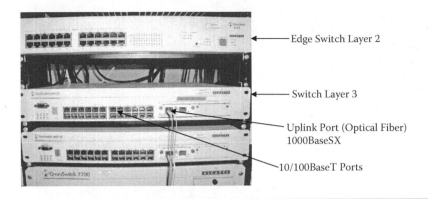

Edge Switch Layer 2

Switch Layer 3

Uplink Port (Optical Fiber)
1000BaseSX

10/100BaseT Ports

Figure 2.5 Edge-core connection with real equipment.

When talking about properly designing a LAN network, one must take into account certain factors such as the type of quality of service desired, the security level overall, and the SLAs (Service Level Agreements). When you want to eliminate the effects caused by collisions in a network, you design a network with switches, but these devices do not eliminate congestion. To solve or at least be aware of the congestion level that could be managed in a controlled way, it is necessary to make a design according to the existing relation from edge to core. What this relation indicates in the worst case would be the maximum that the edge devices could send to the core (as servers, Internet outlets, or WAN outlets; that is, the most critical LAN network devices are located at the core) compared to the total outlet capacity from this edge to the core.

Let us look at an example. Figure 2.6 shows a 10/100BaseT 24-port switch; let us assume that we have an uplink port with 100 Mbps (100BaseF) capacity. The ratio would be given by

Total Capacity to Core (Uplinks): Maximum Transfer to Core

In the worst case, the 24 devices connected to the switch would transmit information to the core, for which the ratio would be

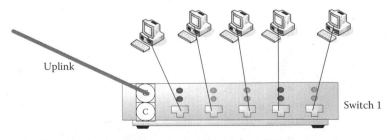

Uplink

C

Switch 1

Figure 2.6 Edge-core relation.

100 Mbps (Uplink Capacity):24 Ports × 100 Mbps (Maximum Transfer Rate)

100 Mbps:2400 Mbps

100 Mbps:2.4 Gbps

That is,

1:24

As we can see, in this case it has been assumed that equipment connected to the switch would be transmitting at maximum capacity, in other words, 100 Mbps, which in practice is not so true; but it is exactly for this reason that a controlled and known congestion ratio is managed so as not to have to buy an over-dimensioned network or to have a congested network.

The ideal ratio, as its name states, would never happen; it would be 1:1. This means that for every device at 100 Mbps there would be a 100 Mbps uplink connection or that for every 10 ports at 100 Mbps there would be one uplink at 1 Gbps or, lastly, that for every 10 connections at 1 Gbps, there would be an uplink at 10 Gbps.

The ratio 1:24 is fairly bad, which in a network with a certain degree of transmission would unchain a severe congestion and consequently a degrading of the services that are being transmitted through such a network.

Manufacturers traditionally recommend a 1:3 ratio or a maximum of 1:5. This means that for every 30 or 50 ports at 100 Mbps in an edge switch, an uplink connection at 1 Gbps would be necessary. In this case, the ratio would be as follows:

For one 10/100BaseT 48-port switch or two 10/10BaseT 24-port switches, stacked or connected via uplink among them and with a 1 Gbps uplink to the core, the ratio would be given by

1 Gbps:48 Ports × 100 Mbps

1 Gbps:4.8 Gbps

1:4.8

The foregoing would mean that for every 10/100BaseT 48-port switch or for every two 10/100BaseT 24-port switches, 1 Gbps uplink would have to be connected. This means that if there are two switches with 48 ports or four switches with 24 ports, in other words, a total of 96 ports at 10/100BaseT, two uplinks of 1 Gbps would have to be connected, performing a trunking function as illustrated in Figure 2.7. The trunking function requires that if we connect two uplinks, the spanning tree (SPT) will have to be activated to prevent formation of a loop between the core and that edge. When the SPT is activated to one of those two uplinks, it will begin working

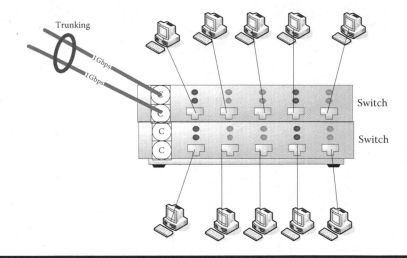

Figure 2.7 Ethernet trunking.

as backup and will be deactivated for data transmission, but when one wants to get the two uplinks to transmit simultaneously, the two must be active. It is because of this that when those two uplinks are configured as trunking, they seem like one only in logical form but with the transmission capacity of the two since the two are physically active. Since the SPT sees only one uplink in logical form, it will not be able to deactivate any of them; consequently, in logical form there would be one uplink with a transmission capacity of 2 Gbps, as shown in Figure 2.7.

The same scheme would occur if instead of having 10/100BaseT ports in the edge, we had 10/100/1000BaseT ports in the edge, because the uplink speeds would be 10 Gbps and the ratio would be exactly the same.

When network analyses have been performed in practice, certain networks have been noted to have good performance and a ratio above 1:5. The only comment I want to make regarding this is that every network has its own transmission rate and ability to transmit simultaneously; therefore, if there is a ratio greater than 1:5, it should be monitored closely in order to detect any increase in conjestion and reduce the ratio when detecting a small degrading of the network, which would begin to affect normal behavior of the services.

To conclude this section, I would like to comment that hubs are good connection devices that have been used for a long time and are inexpensive. But, when designing LAN solutions for services convergence, that is, when in addition to data we are going to transmit voice and video, hubs do not meet the technical characteristics to manage the QoS and we will have to use switches in those connections, and perhaps switches above layer 2.

2.1.2 Core Connectivity

A core consists of one or several switches that essentially receive connections from the edge switches and corporate servers, routers, firewalls, and IP-PBX, among other devices. This means that core devices compose the main part or backbone of the LAN network. Consequently, this device or set of devices must meet very specific and robust characteristics to support all the traffic of the LAN network and the technical demands of the different types of services.

Figure 2.8 shows a typical core connectivity scheme. Here we see that every edge is connected to the core device via the uplink, which in this case is in optical fiber but in certain cases could be in UTP cable. The different servers owned by the company, such as Web server, e-mail server, application server, database server, and file server, could also connect to the core. Since servers are normally located very close to the core switch, it is common that they connect through UTP. Moreover, these devices could connect through optical fiber. It is important to mention, though, that since the servers are used by the majority of people in the organization, these must be located in a strategic place distance-wise and be connected to a high-performance switch; it is for this reason that they must be connected to the core. Even based on the characteristics of the server, it could be that this server has more than one connection to the core switch and more than one way through which the load balancing function can be performed; that is, all connectivity links would be used to transmit information. In case the server is for exclusive departmental use (i.e., for only one department) or the number of simultaneous users is very low, it is possible that even this server is not connected to the core directly but to an edge switch.

It is normal for firewalls (security devices), routers, videoconference stations (although they can also connect at the edge), and IP-PBX to be connected to the core switch. The devices that are performing the function of being the LAN network's core are in a chassis form; that is, the frame arrives and every company selects and installs the type of cards (per slot) required according to their connectivity needs. For example, a core switch must have at least one slot with the management card (there are devices that manage two simultaneously so that in case the main card fails, the second card takes over; these carriers are called *carrier class* due to their availability in case of failure); other slots used by cards with two, four, or eight ports 10GBaseSR (optical fiber multimode to 10 Gbps) for connection of the edge uplinks; cards with 8 or 16 ports 100BaseSX (optical fiber multimode 1 Gbps) for connection of edge uplinks; and cards with 16 or 32 ports 1000/100/10BaseT (UTP cable at 10 Mbps or 100 Mbps or 1 Gbps) for connection of the close edges or for connection of servers, firewalls, IP-PBX, and videoconference stations, among other devices. As a result, when a company is going to design and implement the LAN network, it must select, in addition to the core model (chassis), the type and number of cards for connection to the edges and the other devices that connect directly to the core.

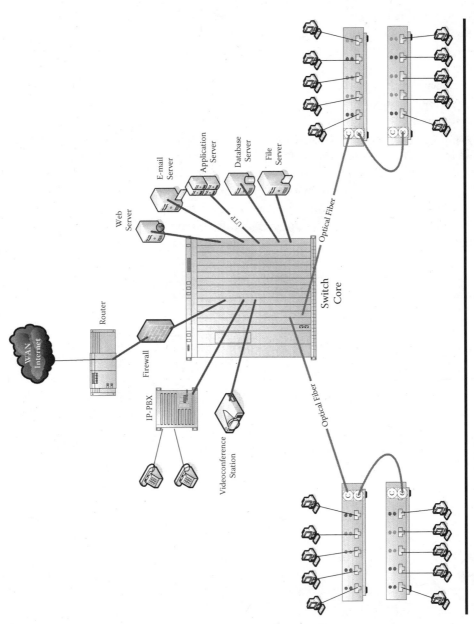

Figure 2.8 Core connectivity.

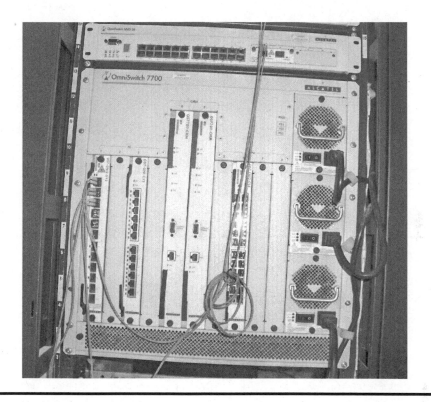

Figure 2.9 Core chassis.

Figure 2.9 illustrates a connectivity with actual devices. This figure shows how the chassis core switch has two management cards (one is redundant in case the principal card fails); one card with availability for 12 ports 1000BaseSX to connect the edge switches; one card with 12 ports 1000BaseT to typically connect the corporate servers, firewall, or edge switches that are found close to the core (≤100 mt) and that the conditions associated to this physical medium allow; and, last, one card with 24 ports 10/100BaseT to connect, for example, IP-PBX, videoconference stations, or important PCs.

2.2 WiFi Solution

This type of wireless solution in LAN networks corresponds to connections on the edge; that is, instead of using some type of guided physical medium, connectivity is possible through this technology without the need of such a guided physical medium. Through WiFi one can connect PCs, IP telephones, Web cams, and printers, among other devices. What matters is that they have a wireless network card. As we saw in Chapter 1, IEEE 802.11 in its different standards specifies operation

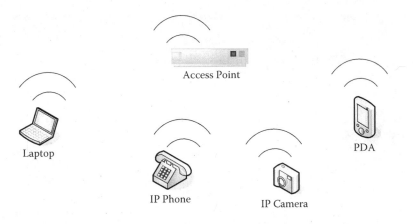

Figure 2.10 Wireless connection.

of this technology; currently on the market is 802.11g with a maximum transmission rate of 54 Mbps and 802.11n is awaited. Figure 2.10 shows a typical wireless connectivity scheme of devices through an access point. Connectivity is also possible without the use of an access point; this case is called an ad hoc network, which is shown in Figure 2.11.

A signal study is necessary when designing a wireless LAN network or at least part of this network, because distances from the access points to the devices and from the different obstacles deteriorate the signal's power and, consequently, the effective transmission rate of the equipment decreases. In the worst cases, there might be disconnection or even failure to connect at all to the wireless network. As illustrated in Figure 2.12, every access point has its own coverage in what is called a *cell*. As seen here, there will be points that are overlapped by more than one access

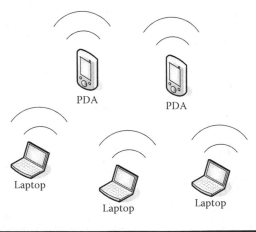

Figure 2.11 Ad hoc network.

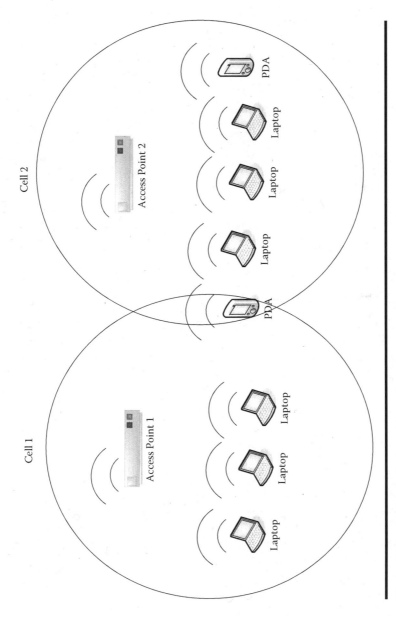

Figure 2.12 Cells of wireless networks.

point, as with the PDA in this figure. The device will connect to the access point with the strongest and most stable signal.

Access points can connect among them in a wireless connection or through Ethernet to reach the edge switches to attain speeds over 54 Mbps.

Figure 2.13 shows both access points connecting to an edge switch through a connection via a UTP cable at 1 Gbps, because every computing device that connects to the access point could transmit up to 54 Mbps for which a 100 Mbps connection would enter the access point and the switch would generate congestion in this section of the network.

This is how information can be transmitted between equipment that is connected through a guided physical medium and equipment connected via wireless and vice versa.

A design that contemplates the cabled part (Ethernet) and the wireless part (WiFi) is shown in Figure 2.14. In this case, we have several access points covering different areas of the company, and these access points connect to the edge switches. It is also possible that certain access points will connect directly to the core switch due to the strategic location of the access points or due to the traffic requirement characteristics of the PCs connecting to such access points.

2.3 LAN Solution with IP

This section discusses a simple IP addressing design because transmission to this section can take place only through switching since the equipment has not been divided into different VLANs (Virtual LANs).

Design with routing will be discussed in the next section.

Before providing a solution with IP addressing, I would like to provide a brief summary of addressing. The intention of this book is not to discuss this theory in detail but to implement it; very good books on this subject are available on the market.

IP addressing in version 4 (IPv4) consists of four numbers in base 10 whose values range between 0 and 255. Every number is 1 byte; therefore, if all bits of this byte are in 00000000 (in base 2), they will correspond to value 0 (in base 10) and if all bits of this byte are in 11111111 (in base 2), they will correspond to value 255 (in base 10). Transmissions in IPv4 can be addressed to one single destination device (unicast), to a group of destination devices (multicast), and to all devices (broadcast). Each of these transmission types, depending on who they are addressed to, will have its own representation in IP addressing.

Every IP address represents the network where the destination equipment to which we want to transmit the information is located; it also represents the specific device in that network that will effectively receive the information. For the end user using the network service this is clear because he will only write something like http://www.domain1.com or ftp serverFTP.domain2.com. Since names (DNS—Domain Name System) are not routable, it is necessary to obtain the IP

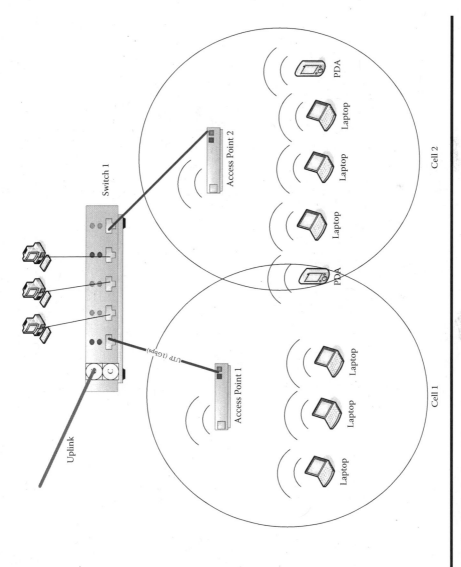

Figure 2.13 Cell wireless integration.

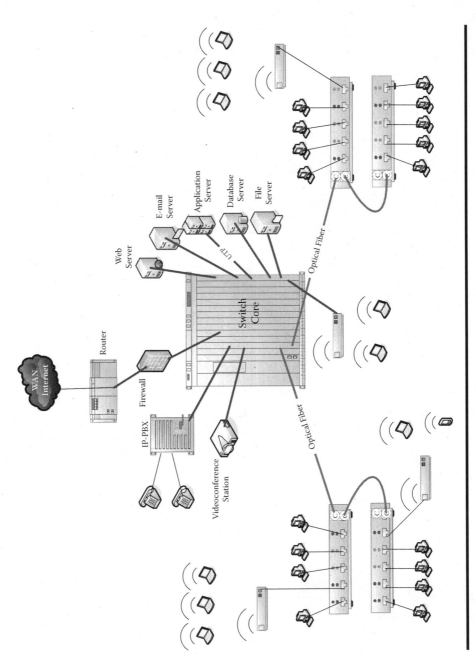

Figure 2.14 LAN design with wireless networks.

address of the equipment with which we want to connect; therefore, the DNS service is required to acquire the IP address. The IP address of device www.domain1.com could be, for example, 200.20.30.1, where the number 200.20.30 (which in network terms would be written as 200.20.30.0) means the network where such Web server is located, and the specific address, .1, that is, 200.20.30.1, is the exact address of the Web server in network 200.20.30.0. If the company has another server, for example, ftp.domain1.com (in practice it could be the same physical device or not; this depends on the configuration made by the network administrator, but let's assume in this example that it is another device with a different IP address) whose IP address is 200.20.30.2, this means that the devices are on the same network (provided no subnets or VLANs are being configured, but this subject has not yet been discussed in this book). If these two devices want to communicate with each other, such communication will be done through switching because they are located on the same IP network and no routing is necessary. Now let's assume that the address of device serverFTP.domain2.com is 200.20.31.1. In this case the network address would be 200.20.31.0 and is, therefore, a network address different from 200.20.30.0. In order for ftp.domain1.com to communicate with serverFTP.domain2.com, a routing function must to take place in order to go from network 200.20.30.1 to network 200.20.31.0. Later on we will explain why addresses 200.20.30.0 and 200.20.31.0 are network identifying addresses.

In IPv4 there are five different ranges, called *classes*, that group the different IP addresses.

Class A addresses are addresses between 1.0.0.0 and 126.0.0.0, where the first number is the network identifier and the next three numbers identify a device within this network. In this case, we know that device 20.0.0.1 belongs to the same network as device 20.20.30.85 and both have 20.0.0.0 as network address. For device 20.0.0.1 to communicate with device 21.0.0.1, it must do routing because they are in different networks.

Class B addresses are addresses between 128.0.0.0 and 191.0.0.0, where the first two numbers identify the network and the following two numbers identify the devices on this network. In this case, we know that device 150.30.0.1 belongs to the same network as device 150.30.80.220. For device 150.30.0.1 to communicate with device 150.31.0.1 it must do routing because they are in different networks.

Class C addresses range from 192.0.0.0 to 223.0.0.0. Here, the first three numbers identify the network and the last number identifies a device within the network. In this case we know that device 200.20.30.1 belongs to the same network as device 200.20.30.220. For device 200.20.30.1 to connect with device 200.20.31.1, it must do routing because they are in different networks.

The three foregoing classes (A, B, and C) are used for unicast transmissions. Class D is the only class used for multicast transmission and this range is between 224.0.0.0 and 239.0.0.0. Class E, which is an experimental class and is not used for regular Internet transmissions, is in the range between 240.0.0.0 and 254.255.255.255.

I will now discuss some special cases.

Network Address. When one has the network address (for example, in Class A 10) and the rest of the bits associated to the devices are in 0 (that is, in Class A 0.0.0 in base 10), it means the network address. In other words, 10.0.0.0 (in bits would be 00001010.00000000.00000000.00000000) is the address for network 10. These types of addresses are very useful for routers when they must perform their routing function. For a Class B, for example, 150.30, the network address would be 150.30.0.0, and for a Class C the network address would be, for example, 200.20.30.0.

Broadcast Address. When one is on a network (for example, in network 10.0.0.0) and wants to send information to all devices on this network, instead of placing all bits associated to devices in 0, one now places them in 1; therefore, the broadcast address for that 10.0.0.0 network will be 10.255.255.255 (in bits it would be 00001010.11111111.11111111.11111111).

Localhost or Loopback Address. Address 127.0.0.0 is not used for network transmissions but to perform local tests, that is, on the same device. For example, if we perform a ping 127.0.0.1, no IP package will be transmitted over the network, but a test will be performed on the IP protocol in the local device.

Now that we have seen some IP concepts, which are enough for this book, we will review what an IP addressing scheme assignment in a LAN network without VLANs management would be like. IP addressing is performed notwithstanding if the LAN connections are being made with Ethernet or WiFi or other LAN solution. For this example I use the IP address of network 10.0.0.0, as mentioned previously, without VLANs management or subnets. Figure 2.15 shows an example of IP address assignment to different devices in the LAN network. Perhaps in real life an addressing scheme like the one shown would reflect the physical location of the device, but according to the scheme shown, it doesn't matter if a routing function must be executed. In fact, every communication that will be done will take place via switching. For example, there are PCs with addresses 10.1.0.2 and 10.2.0.2; laptops with addresses 10.1.0.1, 10.2.0.8, and 10.3.0.1; a PDA with address 10.2.0.1; servers with addresses 10.0.0.4 and 10.0.0.5; a videoconference station with address 10.10.0.1; an IP-PBX (VoIP) with address 10.10.0.2; firewall (security) with address 10.0.0.1; and the core switch to be administered, for example, with address 10.0.0.254. With this connectivity scheme and IP addressing, for example, the PC 10.1.0.2 via Ethernet or the laptop 10.1.0.1 via WiFi or the PDA 10.2.0.1, one can communicate with the e-mail server 10.0.0.5 and the function would be switching.

Figure 2.16 shows the configuration of an IP address of a core switch.

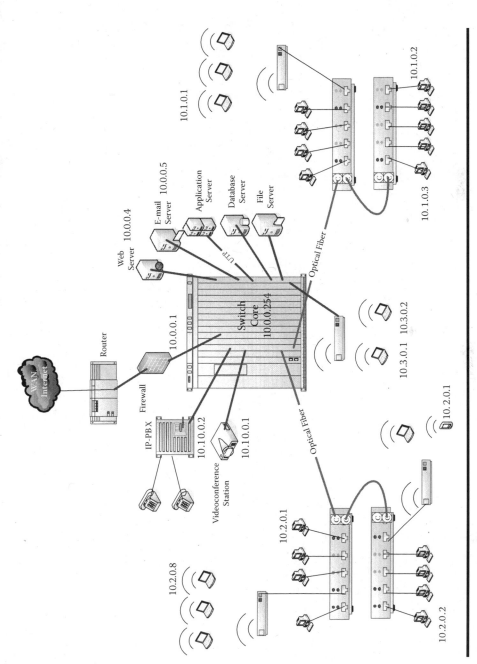

Figure 2.15 LAN design with IP.

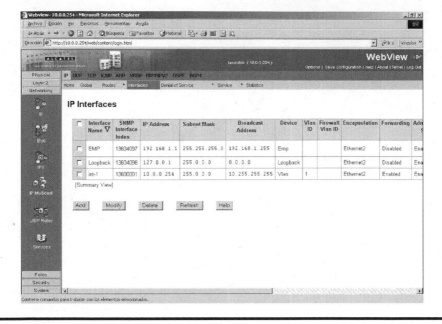

Figure 2.16   IP address in the core switch.

2.4 VLAN Design and LAN Routing with IP

There are several reasons why organizations have implemented VLANs in their different networks. One reason was to create subnets within the same LAN network and then distribute the broadcast in each VLAN, that is, to make sure that one broadcast emitted by a PC will invade only that subnetwork and not the whole organization. Figure 2.17 shows the distribution scheme of a broadcast via VLAN. The consequence of this is that if two devices in different VLANs want to communicate with each other, a routing function must take place, because every VLAN is a different subnetwork. In the example in Figure 2.17, for these two devices that belong to two different VLANs to transfer information, the switch must be layer 3.

Through time manufacturers and users have given the concept of VLANs other interesting uses such as to establish QoS criteria between VLANs or security policies. For example, through VLANs it is possible to create a subnetwork for critical services such as VoIP and isolate this service from the broadcasts in the rest of the LAN network, while providing greater priority to this VLAN service; consequently, VoIP transmission would be assigned greater priority when compared to other services being transmitted over the network.

Under a logical operating scheme between VLANs, when two devices on the same VLAN want to connect with each other, they do so through a switching function, if there are switches connecting them, or through broadcast, if hubs are connecting them. But, if two devices on different VLANs want to connect, they

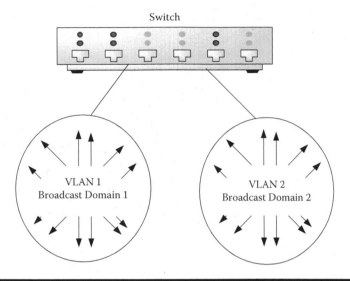

Figure 2.17   Broadcast distributed by VLAN.

must do it through a routing function. It is for this reason that Figure 2.18 shows that broadcast domains belong to each VLAN, but if VLAN 2 wants to communicate with VLAN 3, it must use routing, which is found in a level 3 switch or in a traditional router.

Under a physical perspective, devices may connect at any point on the network without being distributed according to the VLAN to which they belong. Figure 2.19 shows a VLAN scheme in which the physical location of the device is not what matters, but the configuration of the VLAN to which it belongs. For example, VLAN 1 corresponds to the corporate servers. VLAN 2 corresponds to the IP-PBX and the videoconference station; in general, it could be the voice and video services. VLAN 3 corresponds to the PCs connected via Ethernet, and, last, VLAN 4 corresponds to the devices connected in a wireless way via an access point. The foregoing means, therefore, that if a PC in VLAN 3 wants to communicate with a server in VLAN 1, it must route, according to the figure, to the core switch, which is layer 3. If a PC in VLAN 3 sends a broadcast message, this message will reach all devices marked as VLAN 3 and not the devices on other VLANs. For example, it will not reach the servers or the IP-PBX, videoconference station, or devices connected via access points.

We mentioned earlier that every VLAN generates its own broadcast domain, which means that a VLAN by nature is a network or, in a more advanced scheme, could be a subnetwork. If every VLAN is a network, then we could say that VLAN 1, for example, corresponds to network address 10.0.0.0, VLAN 2 to network address 11.0.0.0, VLAN 3 to network address 12.0.0.0, and, finally, VLAN 4 to network address 13.0.0.0. But, in a LAN network design subnets are more

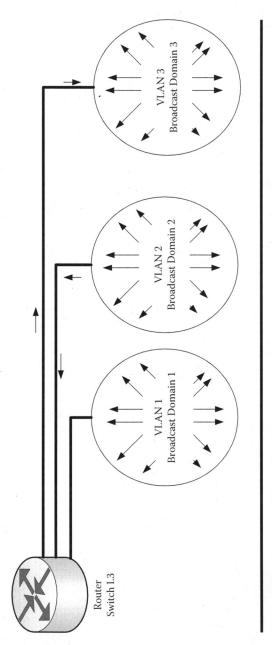

Figure 2.18 Logical scheme of VLANs.

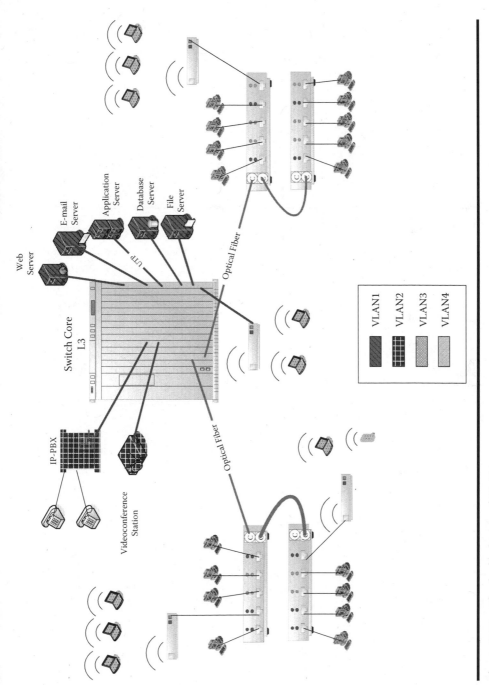

Figure 2.19 Physical scheme of VLANs.

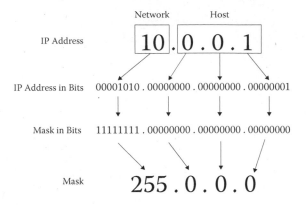

Figure 2.20 Class A addressing.

appropriate and, in fact, are what is implemented. In other words, for example, the complete network address for the LAN corresponds to address 10.0.0.0 and from this address one obtains the subnets.

We will now explain very briefly how to create IP subnets. This book will not go into details of subnetwork management as there are other books that provide this information in detail.

When one configures the IP address, one must also define another parameter called *mask*. The mask tells which bits of the defined IP address correspond to the network address and which bits correspond to the address of a device in the network. On the mask, when a bit is on 1 it means that same bit on the IP address is part of the network address, and when the bit on the mask is on 0 that bit is on the IP address and is part of the equipment address.

For example, in a Class A address the first number identifies the network and the next three numbers identify the device on that network. Then, on the mask, the bits of the first number (byte) will be on 1 and the bits of the mask of the next three numbers (bytes) will be on 0. If all eight bits of the first number of the mask are on 1, then such number corresponds to 255 on base 10. Figure 2.20 shows the mask specification for a Class A address such as 10.0.0.0.

In a Class B IP address, the first two numbers (bytes) will now identify the network and the last two numbers will identify a host within that network. Figure 2.21 shows a specification for a Class B address (150.20.0.0). One can see that for the first two numbers to mean network, a mask 255.255.0.0 must be specified. On the other hand, in an IP Class C address, the first three numbers identify a network and only the last number identifies that network's host. Therefore, the mask for that Class C addressing will be 255.255.255.0, as seen in Figure 2.22.

Now let's assume that we have network address 10.0.0.0 and we want to build subnets within this IP network, for example, to manage VLANs within the LAN network. In this case we must use some bits, which previously meant host, to now

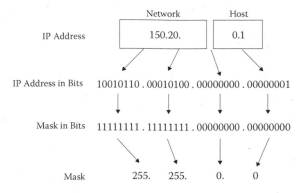

Figure 2.21   Class B addressing.

mean subnetwork. For example, let's assume that we need to create four subnets (VLANs) within our LAN network whose IP address is 10.0.0.0. In order to manage four networks it is necessary to use two bits for the subnets, because the number of subnetworks is obtained from 2^n, where n is the number of bits used to represent the subnets. To solve this case we would use the two most significant bits (first two bits from left to right) in the second number of the IP (byte) address to represent the subnetworks. These bits correspond to bit $2^7 = 128$ and to bit $2^6 = 64$. In this case, by using two bits for subnets one can obtain the following subnetworks associated to the two bits: 00, 01, 10, and 11. Since we are using bits $2^7 = 128$ and $2^6 = 64$, the value of the mask for this subnetwork management will be the sum of the two amounts (128 + 64), that is, 192. Therefore, the mask would be 255.192.0.0. Figure 2.23 shows the mask to create the four subnets both in base 2 (bits) as in base 10 and the four different subnets that result when combining the values of the two bits of the second byte with the values (00, 01, 10, and 11). Another way of representing the mask is through /n where n is exactly the number of bits that are in

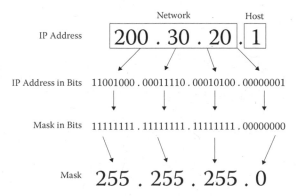

Figure 2.22   Class C addressing.

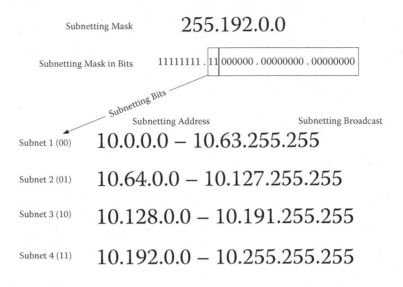

Figure 2.23 Subnets.

1 in the mask. In the foregoing example then it would be /10 because 10 bits are in 1 in the mask. Under this addressing scheme with mask we could have four subnets as we have already mentioned and we could have up to $2^{22} - 2$ in every VLAN. The general formula is $2^n - 2$, where n is the number of bits in the mask that are in 0. Subtract from this value two IP addresses; the first one, for example, in subnetwork 1 (10.0.0.0), which corresponds to the subnetwork address, and the last one (10.63.255.255), which corresponds to the broadcast address in that subnetwork. In this example, $2^{22} - 2$ would be 4,194,302 available addresses for devices. This number, evidently, is too large a number for available addresses for devices, since in reality no VLAN would have this number of devices.

If we apply in logical form the above mentioned subnet scheme to VLANs, such VLANs would look as illustrated in Figure 2.24. Here we are defining only three VLANs to which the following subnet addresses are assigned: 10.0.0.0/10 to VLAN 1, 10.64.0.0/10 to VLAN 2 NS, 10.128.0.0/10 to VLAN 3. The L3 switch (core) would have three IP addresses assigned in this case, one for each VLAN. In VLAN 1 switch L3 has IP address 10.0.0.254, in VLAN 2 it has IP address 10.64.0.254, and in VLAN 3 it has IP address 10.128.0.254. The IP addresses of every VLAN in switch L3 will be the default gateway addresses of the devices associated to every VLAN so that they can route toward another VLAN. For example, there is a PC with address 10.0.0.1 in VLAN 1, whose default gateway is IP address 10.0.0.254, which is exactly the address of switch L3 in VLAN 1. This way PC 10.0.0.1 will be able to communicate with the other VLANs by routing through switch L3. Similarly, there is a PC in VLAN 2 with IP address IP 10.64.0.1, whose default gateway is IP address 10.64.0.254, which corresponds to the address of switch L3 in

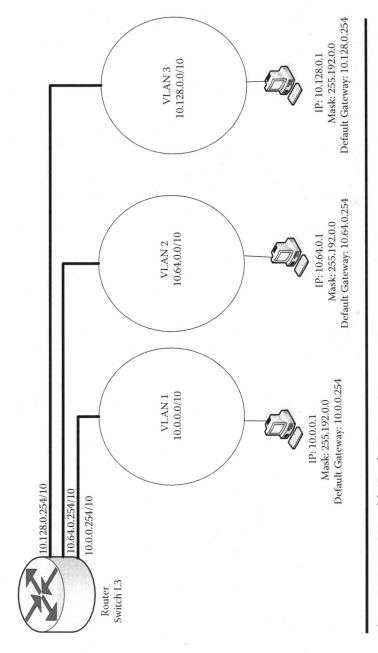

Figure 2.24   VLANs with IP subnets.

VLAN 2. Last, on VLAN 3 there is another PC with IP address 10.128.0.1, whose default gateway is address 10.128.0.1, which corresponds to the address of switch L3 on VLAN 3.

More devices on each of the VLANs could be defined in a similar way configuring IP addresses in the valid range for each VLAN and with their respective mask and appropriate default gateway.

The following is an example of a VLAN design and implementation with a core L3 switch and edge switches in both layer 2 and layer 3 to perform routing functions in the network's edge.

In this example we are using a mask /16, which is 255.255.0.0, to represent the VLANs or subnets. In this case we are using the second byte (8 bits) completely to identify the VLANs. This means that we have the possibility of creating up to 2^8 VLANs, in other words, 256. For every VLAN it would be possible to use the third and fourth byte to represent devices within such VLAN. That is, it could be possible to have up to $2^{16} - 2$ devices, which in this case would be 65,534. To simplify the designs to be shown in this book and the implementations, we will only show the design and configuration of two VLANs; for the rest of the VLANs it would be exactly the same. For this device we have used the first two IP subnets (10.0.0.0/16 and 10.1.0.0/16). The main switch, which is layer 3, has the two VLANs configured. VLAN 1 has IP address 10.0.0.254/16, and VLAN 2 has IP address10.1.0.254/16. This core switch has one server connected in VLAN 1 and two servers in VLAN 2. Edge switch 1, which is also layer 3, has the two VLANs configured. VLAN 1 has IP address 10.0.0.253/16, and VLAN 2 has IP address 10.1.0.253/16. Two PCs are connected to each of these VLANs in edge 1 switch. Edge switch 2, which is only layer 2 and therefore cannot perform routing, has two VLANs configured. VLAN 1 has IP address 10.0.0.252/16 and VLAN 2 has IP address 10.1.0.252/16. As in edge 1, two PCs are connected to each of these VLANs in edge 2.

As a first step in the analysis of this design, we will look at the core switch. There is one server connected to VLAN 1 with IP address 10.0.1.1/16, which shows that it belongs to IP subnet 10.0.0.0/16 associated to VLAN 1. This server has the core switch in VLAN 1 as default gateway, in other words, to IP address 10.0.0.254. Two servers and an IP-PBX, whose addresses are 10.1.1.1/16, 10.1.1.2/16 and 10.1.1.3/16, are connected to VLAN 2, showing that it belongs to subnet 10.1.0.0/16, associated to VLAN 2. These servers have the switch core on VLAN 2, which is IP address 10.1.0.254, as default gateway.

For our analysis, let's first look at edge 1, which is a layer 3 switch and therefore cannot perform the routing function either. Two PCs with IP address 10.0.0.1/16 and 10.0.0.2/16, respectively, are connected in VLAN 1; this shows that they belong to IP subnet 10.0.0.0/16 associated to VLAN 1. The default gateway of these PCs is their edge switch or edge 1 in VLAN 1, which corresponds to IP address 10.0.0.253. Two PCs whose IP addresses are 10.1.0.1/16 and 10.1.0.2/16, respectively, are connected to VLAN 2, which demonstrates that they belong to subnet

10.1.0.0/16 associated to VLAN 2. The default gateway of these PCs is their edge switch or edge 1 in VLAN 2, which corresponds to IP address 10.1.0.253.

Edge 2 is a layer 2 switch and therefore will not perform routing. Consequently, devices connected to this switch will have to rely on the closest layer 3 switch. For this case, the closest switch is precisely the core switch. Two PCs, whose IP addresses are 10.0.10.1/16 and 10.0.10.2/16, respectively, are connected in VLAN 1, which demonstrates that they belong to IP subnet 10.0.0.0/16 associated to VLAN 1. These PCs have the switch core in VLAN 1 as default gateway, which corresponds to IP address 10.0.0.254. To VLAN 2 are connected two PCs with IP addresses 10.1.50.1/16 and 10.1.50.2/16, respectively, which demonstrates that they belong to subnet 10.1.0.0/16 associated to VLAN 2. The default gateway of these servers is the core switch in VLAN 2, which corresponds to IP address 10.1.0.254.

Figure 2.25 shows the physical implementation design of VLANs. Under the design shown in Figure 2.25 we must analyze connectivity of the uplink ports (optical fiber) of edge switches (edge 1 and edge 2) to the core switch. If we review the figure we can see that uplink ports in the edge switches and the core switch belong to VLAN 1. If we want to transmit from PC 10.1.0.1 located on edge 1 to server 10.1.1.1 located on the core or to PC 10.1.50.1 located on edge 2, this transmission must be made through switching. But, a problem arises here: The ports (uplinks and downlinks) that communicate these switches belong to VLAN 1. Therefore, devices in VLAN 2 located in different switches are incommunicated and are isolated without communication. To solve this problem, standard 802.1q was defined; this standard is configured on the ports through which one wants the packages from other VLANs to be transmitted. Since uplink ports of edge switches and downlink ports of the core switch belong to VLAN 1, it is necessary to configure 802.1q in these ports and then the segments generated by devices in VLAN 2 can be transmitted through these ports. Standard 802.1q is a mark shown on Figure 2.26.

Figure 2.27 shows the logical design of VLANs and the connected PCs with their respective default gateway. As we can see, the default gateway in VLAN 1 for some devices is address 10.0.0.254 and for others it is address 10.0.0.253, depending on which switch they are connected to. If they are connected directly to the core switch or to edge 2, which is layer 2, the default gateway is 10.0.0.254, as can be seen in Figure 2.25. For devices connected to edge 1 in VLAN 1, the default gateway address would be 10.0.0.253, since this switch is layer 3. As a consequence of this logical design, we can see that no matter the physical location of the device, all PCs belonging to the same VLAN would transmit information via switching (in case they are connected through switches. as is the case in this design). The same situation is observed for VLAN 2.

The following figures show examples of configurations made on certain devices. Figure 2.28 shows two VLANs configured in the core switch (10.0.0.254).

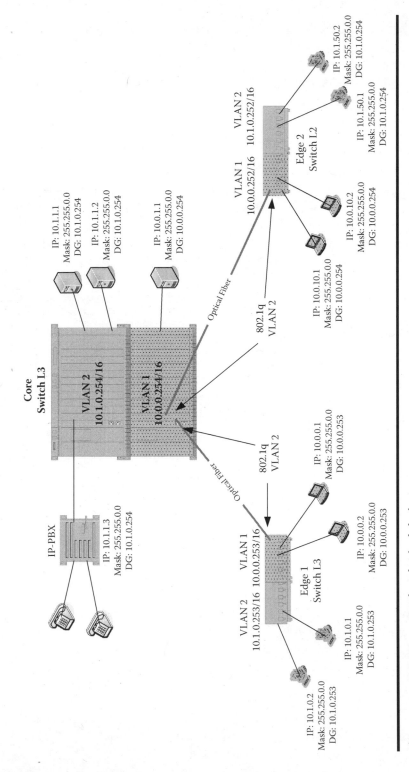

Figure 2.25 VLANs example (physical design).

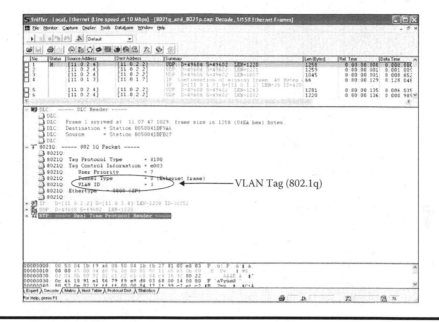

Figure 2.26   Frame with 802.1q.

In Figure 2.29 we see the IP addresses of the two VLANs. VLAN 1 with IP address 10.0.0.254, mask 255.255.0.0, and broadcast address 10.0.255.255. VLAN 2 with IP address 10.1.0.254, mask 255.255.0.0, and broadcast address 10.1.255.255.

Figure 2.30 shows the configuration of 802.1q in the core switch. This configuration must also be performed at the edge switches.

Last, Figure 2.31 shows how the Routing Information Protocol (RIP) has been activated.

It is well known that a new IP protocol has been specified (IPv6). We are currently waiting for carriers and Internet service providers (ISPs) to make it a reality in Internet operation since all devices manufactured during recent years fully support it. Next, we will discuss certain important aspects of IPv6.

One of the most interesting frontiers in Internet research and development is the problem posed by new applications, created often, that demand facilities or characteristics that current protocols cannot provide. For example, multimedia demands protocols for efficient transportation of images and sounds. Moreover, real-time audio and video communications have generated a demand for protocols that can guarantee submittal of information within a maximum delay, and for protocols that can synchronize audio and video chains. Internet has experienced several years of sustained exponential growth, doubling its size every nine months or less. In early 1994, a new host appeared every 30 seconds and the rate has increased considerably. Surprisingly, Internet traffic has grown faster than the number of networks. Traffic increase can be attributed to any of the following reasons: First,

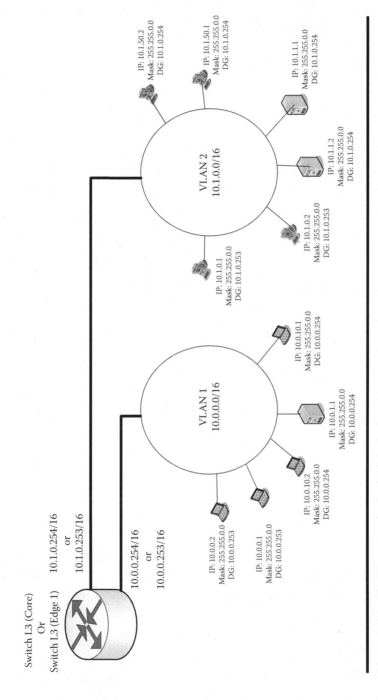

Figure 2.27 VLANs example (logical design).

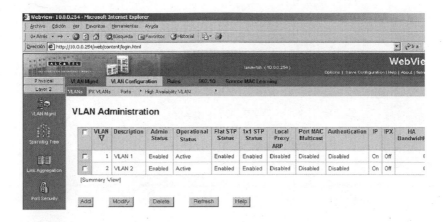

Figure 2.28   VLAN configuration.

the Internet population is moving from academicians and scientists to the general public, thus giving rise to the new use of Internet as a hobby during leisure hours. Second, new applications that transfer images and videos in real time generate more traffic than those containing text only. Third, automated search engines generate substantial amounts of traffic while they exhaustively search every place of the Internet for the information requested. In its expansion to new industries and countries, the Internet is fundamentally changing, acquiring new administrative authorities. Changes in authorities cause changes in administrative policies and new mechanisms are designed to implement such policies. As we have seen, both the Internet architecture and the protocols that it uses evolve separately from the

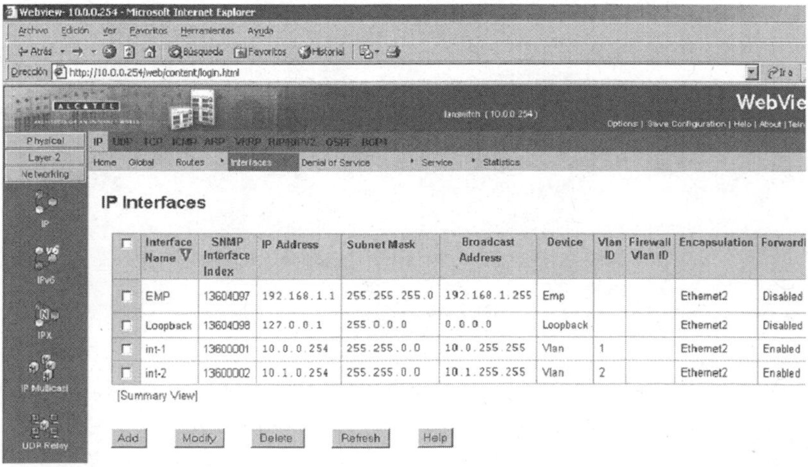

Figure 2.29   VLAN IP address.

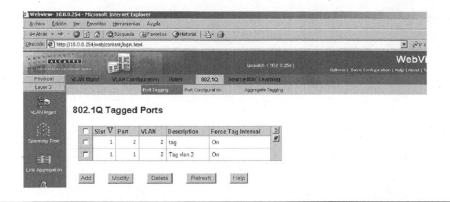

Figure 2.30   VLAN IP address.

idea of a centralized model. Evolution continues as new national centers adhere to the network, resulting in complex regulation policies. When multiple corporations interconnect private networks using TCP/IP through the Internet, they face similar problems in their intent to define interaction policies and find mechanisms to implement such policies. Consequently, much of the research and engineering efforts around TCP/IP are aimed at searching the means to accommodate new administrative groups. The purpose of protocol IPv6 maintains many of the characteristics that contributed to the success of IPv4. In fact, designers define IPv6 like IPv4 and certain new modifications. For example, IPv6 still supports sending without connection (i.e., allows independent routing for each datagram), allows sender to select the size of the datagram, and demands the maximum number of hops that a datagram can make before being discarded. As we will see, IPv6 also

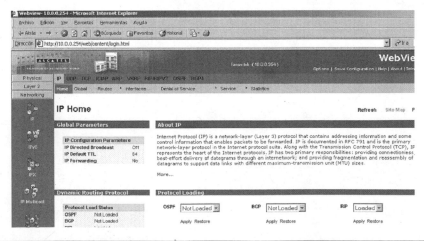

Figure 2.31   Routing function.

maintains most of the concepts implemented by the options of IPv4, including the fragmentation facility and source routing.

Despite the many similarities in concept, IPv6 changes most of the protocol details. For instance, IPv6 uses longer addresses and adds some new characteristics. More importantly, IPv6 fully reviews the datagram format, replacing the length option of the IPv4 package for a series of fixed format headers. We will look at the details after considering the major changes and the reasons for such changes.

The changes introduced by IPv6 can be grouped in the following five categories:

Longer Addresses. The new address length is the most notable change. IPv6 quadruples the size of IPv4 addresses from 32 bits to 128 bits. The available space for an IPv6 address is so large that it avoids the potential exhaustion of addresses.

Flexible Header Formats. IPv6 uses a new incompatible datagram format. Contrary to IPv4, which uses a fixed format datagram header where all fields except the option occupy a fixed octet number, IPv6 uses an optional set of headers.

Improved Options. Like IPv4, IPv6 allows datagrams to include optional control information. IPv6 includes new options that provide additional facilities not available in IPv4.

Support for Resource Reservations. IPv6 replaces the IPv4 type of resource specification with a mechanism that allows prior reservation of network resources. Specifically, the new mechanism supports applications such as real-time video, which requires guaranteed bandwidth and delay.

Provision for Protocol Extensions. Perhaps the most significant change in IPv6 is the deployment of a protocol where all resources are fully specified to another and then additional characteristics are allowed. The extension capacity has the potential to allow the IETF to adapt network hardware changes or new applications to the protocol.

IPv6 completely changes the datagram format. A datagram has a base header with a fixed length, followed by zero or more extension headers, followed by the data. IPv6 has the capacity to address longer lengths than IPv4; therefore, the base header contains more information in IPv6 than in IPv4. The options and certain fixed fields that appeared in the IP4 datagram header must be moved to the extensions of the headers in IPv6. Overall, the changes in datagram headers are as follows:

The alignment has changed from 32-bit multiples to 64-bit multiples.

The header length field has been eliminated, and the datagram length field has been replaced by a field called PAYLOAD LENGTH.

Length of the origin and destination addresses fields have increased to 16 octets each.

Fragmentation information has been moved outside the fixed fields in the header base for an extension of the header.

The TIME-TO-LIVE field has been replaced by a field called HOP LIMIT.

The TYPE OF SERVICE field has been replaced by a field called FLOW LABEL.

The PROTOCOL field has been replaced by a field that specifies the type of the following header.

Many fields in an IPv6 base header correspond directly to a field in the IPv4 header. As in IPv4, the first four bits (VERS field) specify the protocol version; the VERS field in IPv6 always contains a 6 in the IPv6 datagram.

As in IPv4, the ORIGIN ADDRESS and DESTINATION ADDRESS fields specify the source address and the destination address. In IPv6, however, every address requires 16 octets.

The field HOP LIMIT corresponds to field TIME-TO-LIVE in IPv4. Unlike IPv4, when we interpret a time-to-live as a combination of hops (hop-count) and maximum time (timeout), IPv6 interprets the value as the maximum number of hops that a datagram can perform before being discarded.

The datagram manager in IPv6 specifies lengths in a new way. First, because the header size is fixed at 40 octets, the base header does not include a field for the header's length. Second, IPv6 replaces the length field in IPv4 for a 16-bit field called PAYLOAD LENGTH, which specifies the number of octets carried in the datagram, excluding the datagram header. An IPv6 datagram can have up to 64K data octets.

A new mechanism in IPv6 supports reservation of resources and allows routers to associate every datagram with an assigned resource. This consists of a route through a lengthwise Internet in which the intermediate router guarantees a specific service quality. For example, two applications that need to send video can establish a flow in which the delay and bandwidth are guaranteed. Conversely, in a network a user may need to specify the quality of service desired; then, he can use a flow to limit traffic in a specific computer or in a specific application sent.

The FLOW LABEL field in the header base contains information that routers use to associate a datagram with a specific flow and priority. The field is divided into two subfields.

With FLOW LABEL the first four bits of subfield TCLASS specify the traffic for the datagram. Values from 0 (zero) to 7 (seven) are used to specify the time-sensitivity of flow-controlled traffic; values from 8 to 15 are used to specify a priority for nonflow traffic. The remaining 24 bits contain the FLOW IDENTIFIER field. The origin selects a flow identifier when it establishes the flow. No potential conflict exists between computers because a router uses the combination of the origin destination of the datagram and the flow identifier when it associates a datagram with a specific flow.

The paradigm of fixed base header followed by a set of extension headers was selected as a commitment between generalizing and efficiency. To be completely general, IPv6 needs to include function-supporting mechanisms such as fragmentation, origin routing, and authentication. However, choosing to allocate fixed fields in the datagram header for all mechanisms is inefficient because many datagrams do not use all the mechanisms. For example, when we send a datagram through a simple local area network, a header containing fields of empty addresses may occupy a substantial fraction of every frame. Paradigms of extension headers in

IPv6 work in a similar way to the options in IPv4—a delivering team can choose which extension headers to include in a datagram and which to omit, which means that extension headers provide maximum flexibility.

Consequently, extension headers in IPv6 are similar to the options in IPv4, with the difference that every datagram includes the extension headers only for those facilities that the datagram will use. Each of the headers, whether base or extension, contains a field called NEXT HEADER. The software in an intermediate router and in the final destination where the datagram is to be processed has to use the value found in the NEXT HEADER field of each header to parse the datagram. To fully extract the information from a datagram header in IPv6 requires a sequential search through the headers. Note: The NEXT HEADER field in every header specifies the type of the following header.

It is true that parsing an IPv6 datagram that has only one base header and data is as efficient as parsing a datagram in IPv4. In addition, we will see that intermediate routers rarely need to process all extension headers.

As in IPv4, reassembly of the datagram in IPv6 takes place at the final destination. However, designers made an unusual decision about fragmentation. Remember that IPv4 requires an intermediate router to fragment any datagram that is too large for the MTU (maximum transfer unit) of the network over which they will be carried. In IPv6, fragmentation is restricted to the source node. Before sending a message to the network, it is necessary to discover the network MTU, a technique to find out the minimum MTU along the path toward the message destination. Before the datagram is sent, it is fragmented at its place of origin so that every datagram is smaller than the MTU of the route. Therefore, no intermediate fragmentation by a router is necessary.

The base header of IPv6 does not contain analog fields to the fields used for fragmentation of a header in IPv4. Instead, when fragmentation is required, the origin point inserts a small extension header after the base header in each fragment. IPv6 retains a large part of the fragmentation performed in IPv4. Every fragment must be a multiple of 8 octets; one bit in the MF field marks the last fragment like bit MORE FRAGMENTS in IPv4; and the identification field of the datagram carries the only ID that the receiver uses to assemble fragments.

The reason for using end-to-end fragmentation is the possibility of reducing router overcharges, allowing them to manage a larger number of datagrams per time unit. Actually, the CPU overcharge required for fragmentation IPv4 could be significant—in a conventional router CPUs may reach 100% use if they fragment almost all the datagrams they receive. However, end-to-end fragmentation has an important consequence: it modifies an essential assumption about Internet.

To understand the consequence of end-to-end fragmentation, remember that IPv4 was designed to allow changing the routers at any time. For example, if a network or a router fails, traffic may be routed along a different path. The main advantage of a system like this is its flexibility—direction of traffic may be modified without having to cut communication and without notifying the data source or destination. In IPv6, however, the pathway cannot be changed as easily because a change in route implies a change in MTU. If MTU of the new pathway is smaller than the original pathway, either the datagram must be fragmented in the router or the source point of the data must be informed. The problem can be summarized this way: An Internet protocol that uses end-to-end fragmentation needs the source to discover the MTU of the pathways for each destination, and then it fragments all datagrams being sent that are larger than the MTU of the respective pathway. End-to-end fragmentation does not accommodate to changes along the way. To solve the problem of changes along the way that affect the pathway's MTU, IPv6 allows intermediate routers to perform IPv6 tunneling through IPv6. When an intermediate router needs to fragment a datagram, the router does not insert an extension header in the fragment, nor does it make changes in the base header. Instead, the intermediate router creates a new datagram that encapsulates the original as data. The router divides the new datagram into fragments, placing the base header in each one and inserting a fragment extension header in each one. Finally, the router sends every datagram to its final destination. At the final destination the original datagram can be formed by assembling the fragments in a datagram, extracting the data portion from the result.

Headers have a list of addresses that specify intermediate routers through which the datagram must pass. The NUM_ADDRS field specifies the total number of addresses in the list and the NEXT_ADDRES field specifies the following address to which the datagram must be sent.

In IPv6, every address occupies 16 octets, four times the size of the IPv4 address. The size of the address guarantees that IPv6 can tolerate any reasonable addressing scheme. In fact, if a designer wants to later change the addressing scheme, the space of the address is large enough to accommodate the new addressing. The size of the space for IPv6 addressing is difficult to imagine. One way to look at it is relating it to the size of the population: the address space is large enough so that every person on the planet can have enough addresses to set up his own Internet with the same size as that of the current Internet. Another way to look at it is related to deple-tion of addresses. For example, consider how much time it would take to assign all addresses. From a complete 16 octets one can obtain 2^{128} values. Therefore, the address space is larger than $3.4*10^{38}$. If addresses were assigned at an average of one million addresses every microsecond, it would take around 20 years to assign all possible addresses.

Although it solves all capacity problems, the size of the new address implies new problems: people who maintain Internets must read, input, and manipulate such addresses. Obviously, the binary notation is unsustainable. However, the dotted decimal notation used by IPv4 does not make these addresses sufficiently compact either. To understand why, consider an example of a number of 128 bits in dotted decimal notation:

104.230.140.100.255.255.255.255.0.0.17.128.150.10.255.255

To help build more compact, easier-to-type addresses, IPv6 designers recommended the use of two-point hexadecimal notation (abbreviated as colon hex), in which the value of every group of 16 bits is represented in its respective hexadecimal notation and is separated by a colon. For example, the above stated value in dotted decimal notation transferred to colon hex converts to

68E6:8C64:FFFF:FFFF 0:1180:96A:FFFF

The obvious advantage of colon hex is that it requires fewer digits and less separation characters than dotted decimal notation. Moreover, colon hex notation includes two techniques that make it even more useful. First, it allows suppression of zeros; a chain of contiguous zeros is replaced by a colon. For example, the address

FF05:0:0:0:0:0:0:B3

may be written as

FF05::B3

To make sure that compression of zeros does not result in ambiguous addresses, the proposal requires that compression be used only once within an address. The compression of zeros is especially useful when used with the proposed address assignment scheme since many addresses will contain chains of contiguous zeros. Second, colon hex notation incorporates decimal suffixes with dots; as we will see, this property is to be used during the transfer from IPv4 to IPv6. For example, the following is a valid chain within the colon hex notation:

0:0:0:0:0:0:128.10.2.1

Note that although the numbers separated by colons specify a value of a 16 bits amount, the numbers within the dotted decimal position represent an octet. With the foregoing chain, as well, the compression of zeroes can also be applied, resulting in a chain similar to an IPv4 address:

::128.10.2.1

2.5 LAN-MAN Connection

Last in this chapter I would like to explain the connection to the MAN network access via last mile. Chapter 3 will explain the design of carrier networks, that is, last miles and core structure.

As can be seen in Figures 2.8 and 2.25, the core switch is part of the network main structure to which are connected edge switches and also devices such as routers or firewalls. In this case, we will show a connection of the LAN network from the core switch to the carrier network, for which a router will be needed. Direct connection from the core switch is also possible.

Due to the diversity of layer 1 and layer 2 technologies by means of which a MAN or WAN connection can be made, it is necessary to show different scenarios of possible connection types.

A first group of possible solutions for the last mile would be those where technologies such as xDSL (HDSL, VDSL, ADSL) or line T1, E1, T3, or E3 are used as layer 1. For these types of layer 1 technologies it is possible to use PPP and HDLC as layer 2 in the case of end-to-end communications, which are commonly called *clear channels*, or technologies that perform switching such as Frame Relay, ATM, or MPLS, although MPLS currently is a technology associated to the carrier network's core. As explained in Chapter 1, a last-mile connection could be made through PPP/HDSL or HDLC/HDSL or PPP/E1 or HDLC/T1 or Frame Relay/HDSL or ATM/ADSL. In order to be able to make this type of connection, a device with at least one port with this type of technology (switch or router) is needed. If the core switch does not have this option, then a router must be used to supply this type of MAN or WAN connection in the way shown in Figure 2.32.

Figure 2.32 shows how the connection router to MAN or WAN connects to the core switch through Ethernet and how the connection router through some of the technologies shown can connect to the carrier.

With the layer 1 technologies described in this section we can transmit up to the following rates: ADSL upstream 1 Mbps and downstream 8 Mbps, ADSL2 upstream 1 Mbps and downstream 12 Mbps, ADSL2+ upstream 2 Mbps and downstream 24 Mbps, HDSL T1 (1.5 Mbps) or E1 (2 Mbps) symmetrically, VDSL 52 Mbps theoretical, and VDSL2 theoretical with 250 Mbps (although at 500 mt it shows a rate of 100 Mbps and T3 at 44 Mbps).

Figure 2.33 shows an actual device with this type of connection.

Another type of connection that can take place through last mile is through high-speed wireless schemes such as LAN Extended (802.11) through point-to-point or point-to-multipoint antennae, which can transmit over several kilometers, or through WiMAX (802.16). In this case connection is direct from bridge 802.11 or 802.16 to the LAN network core. With this technology 10 Mbps or more can be attained. In fact, WiMAX has been specified until present for approximately 75 Mbps, although in many cases this transmission rate has not been attained. Figure 2.34 shows the connection scheme through this type of technology.

Figure 2.35 shows an actual device connecting the last mile through LAN Extended.

When a company requires a last-mile connection with greater speed than those previously mentioned, it is possible to use another type of layer 1 technology. In this case we would be referring to SONET, SDH, or Ethernet. As a second case, we

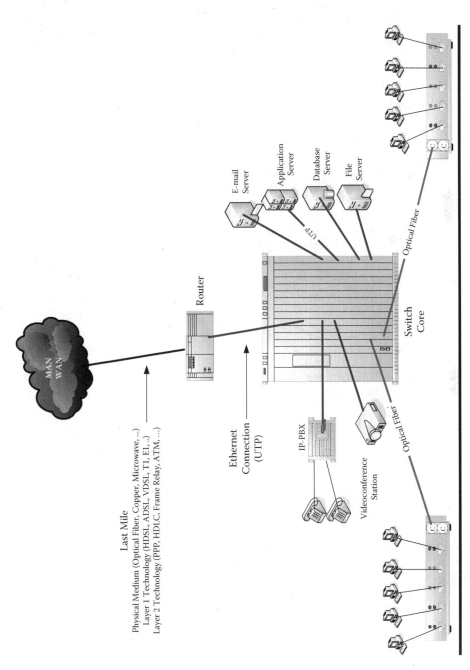

Figure 2.32   LAN-MAN connection case 1.

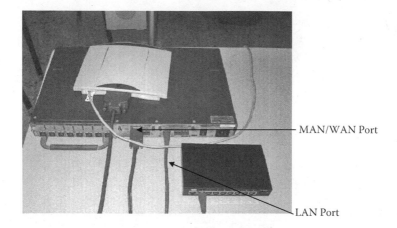

MAN/WAN Port

LAN Port

Figure 2.33   LAN-MAN connection case 1 with equipment.

describe connection through SONET or SDH. Normally, LAN core switches do not support SONET or SDH ports, and therefore one must connect to the LAN core switch a switch with at least one SONET or SDH port to connect it to the MAN on the last mile. If the core switch supports SONET or SDH ports, connection can be direct. Figure 2.36 shows a switch with SONET or SDH ports, which is the frontier device with which the last-mile connection is made. With SONET or SDH it is possible to connect in the last mile, for example, at 155 Mbps through an OC-3 port in SONET or STM1 in SDH, and at 622 Mbps through an OC-12 port in SONET or STM-4 in SDH. Speeds higher than these are not used traditionally in the last mile but are part of the transmission rates from the core to the MAN or WAN network.

Figure 2.37 shows an actual device with this type of connection for access to last mile by means of SONET or SDH.

Finally, another type of high-speed access is through Ethernet. In this case, the same LAN network core device is used to make the last-mile connection in the last mile to the carrier. Due to the distance that separates the LAN network from the carrier, it is normal for the type of optical fiber port and the optical fiber itself used to be different from the one used to make the LAN connections. Figure 2.38 shows a last-mile connection scheme through Ethernet.

If VLANs are defined in the LAN network, which happens often, the router connecting to the last mile would be part of one of such VLANs. Traditionally, it is associated to the network administration VLAN, but it can also be located in a different VLAN because security mechanisms can be applied to control access to the LAN network through the last mile. Therefore, to transfer the information to the last mile, a routing function must take place. Figure 2.39 shows a design where the router connecting to the last mile is connected to the LAN network through a separate VLAN (VLAN 3). This case shows a private IP addressing scheme.

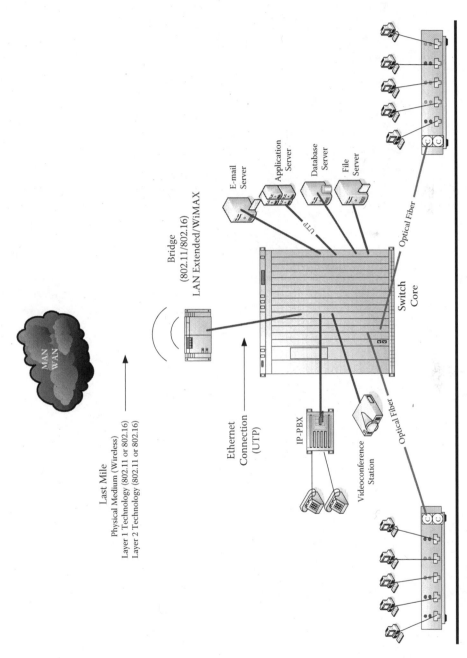

Figure 2.34 LAN-MAN connection case 2.

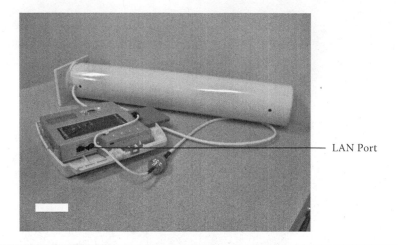

LAN Port

Figure 2.35   LAN-MAN connection case 2 with equipment.

Last, Figure 2.40 shows the connection and routing logical scheme for VLAN 3, through which connectivity to the last mile is attained. As we can see, switch L3 is the main router within the LAN network; therefore, the router as such is used to perform routing toward MAN or WAN.

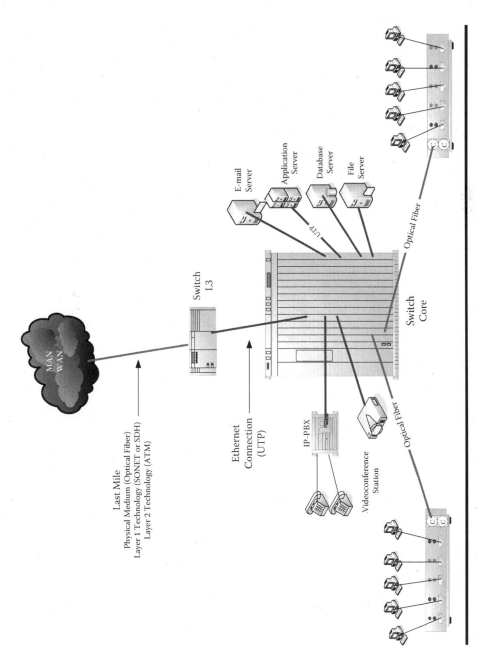

Figure 2.36 LAN-MAN connection case 3.

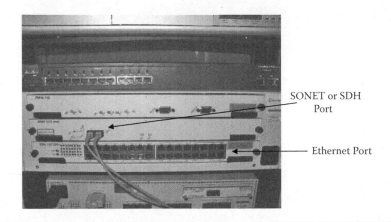

Figure 2.37 LAN-MAN connection case 3 with equipment.

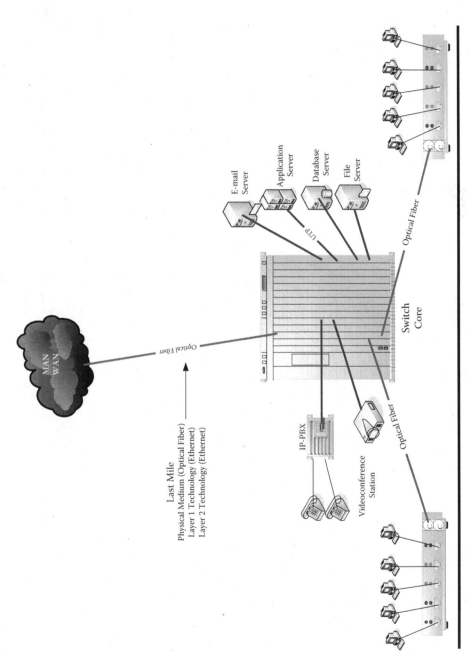

Figure 2.38 LAN-MAN connection case 4.

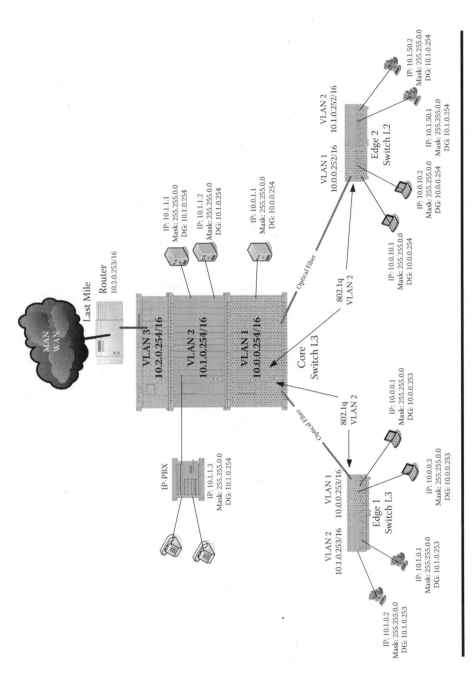

Figure 2.39 Router connection with VLANs (physical design).

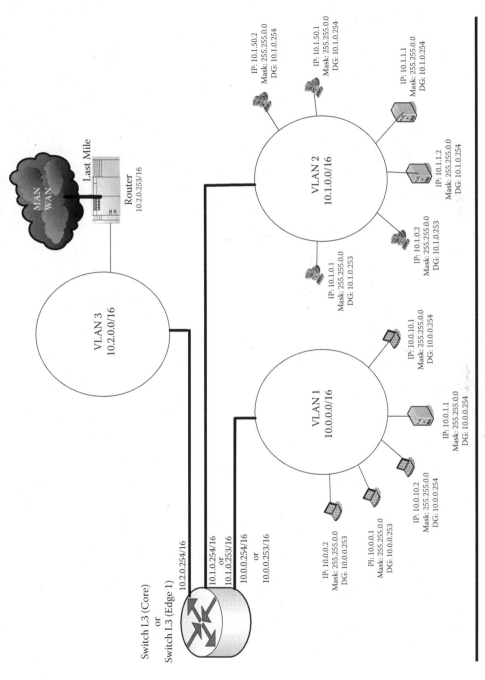

Figure 2.40 Router connection with VLANs (logical design).

Chapter 3

MAN/WAN Network Design

This chapter deals with the designs and configurations that cover carrier MAN or WAN networks depending on whether they cover a city, a region, or a country, or if international connections are made. The chapter has been divided into two main sections to tackle this topic. The first part is related to technologies, forms of connectivity, and access-network-related designs. In other words, this refers to the networks connecting companies and even individual users to the carrier networks. This type of connection is also known as *last mile, last kilometer, first mile*, and *first kilometer*. The second section deals with technologies, the connectivity forms, and the designs related to the core of the carrier networks. In other words, it deals with the networks in charge of transporting the entire carrier's flow. The objective of this book is not to present an exhaustive analysis of every single technology, but to present the most relevant ones. It is likely that other technologies that are not being analyzed follow the same model as those analyzed in this book. Two important topics that correspond to GMPLS networks are analyzed at the end of this chapter, which tackles the MPLS future version and therefore carrier networks and IP addressing design related to customers' network interconnection through the carrier networks.

3.1 Last-Mile Solution

This first section complies with the objective of connecting customers' LAN networks in the closer connection toward the carrier networks, which was the last topic dealt with in the previous chapter. The last chapter analyzed the connection perspective on the customer's side and this chapter covers the perspective on the carrier's side, so as to have a congruent and supplementary idea between the carrier and the customer.

Figure 3.1 shows the connectivity part that corresponds to access or last-mile networks.

3.1.1 LAN Extended

Some operators have suggested, as a first solution, the use of the 802.11 technology to have access to the carrier. Point-to-point or point-to-multipoint antennae are used with the multipoint on the carrier's side. The traditional scheme of these antennae is that they are connected toward the bridge in an Ethernet port. Figure 3.2 shows the equipment giving a LAN Extended solution.

Figure 3.3 shows the physical connectivity scheme in the last minute through LAN Extended. By means of this type of connection, the switch core of the LAN network is connected to the bridge by means of a port in Ethernet with UTP, except if the distance is greater than the one supported by this type of cable. The bridge is connected to the antenna by means of a coaxial cable and this way, having a line of sight with the antenna located in one of the carrier's sites, the physical connection will be made as in the customer's LAN network. The objective in the carrier's inlet consists of being able to connect to the carrier's backbone. No matter what the connection form, this equipment receiving the bridge in the LAN Extended solution will need to have available an Ethernet port.

At present there is an inconvenience with this type of suggested connection, which is the broadcast management. The reason is that the equipment used can only be layer 2, in which case there would not be a way of dividing into sectors the broadcast at the LAN network level so as not to pass to the last mile. The problem is that if VLANs are shaped, the packets cannot be routed, and therefore the transmission toward the other edge through the last mile is not possible. The above mentioned statement suggests having at least one layer 3 equipment, which could become the core switch or the LAN Extended bridge; in other words, the important thing is to divide the broadcast into sectors through the different broadcast domains, that is, VLANs. Figure 3.4 defines two VLANs. The first VLAN is the internal one, but in reality there could be several internal VLANs, as designed in Chapter 2. The second VLAN is the external one, where the LAN Extended bridge and the connection to the carrier are found. The passage of the broadcast from the LAN network to the last mile would be blocked, and in order to transmit data toward a connection or an external edge of the LAN network it should be routed in the switch of core layer 3. Likewise, the layer 3 equipment could be the one usually working as a connection bridge in the LAN Extended.

Another connection could be that of making a point-to-point communication between two sites of an organization, in which case the carrier's network would

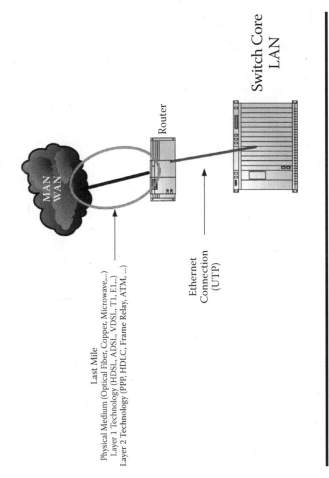

Figure 3.1   Access network.

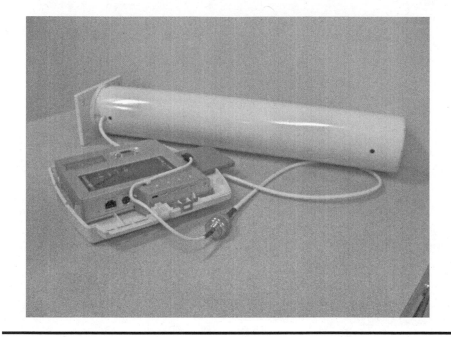

Figure 3.2 LAN Extended (antenna and bridge).

not be used. This type of solution is broadly used, even though this is not the main study objective of this book. As in the previous case in which a carrier is used, the implementation of VLANs is recommended to avoid the broadcast of a LAN network to propagate to the other LAN by means of a LAN Extended. This scheme is shown in Figure 3.5.

3.1.2 Clear Channel

Another way of connecting in the last mile with the carrier, one that is widely used, is the one commercially known as *clear channel*. It is a point-to-point type of connection between the customer's LAN network and the carrier's inlet point. It is a scheme similar to that of the LAN Extended, only that the type of solution involves technologies other than 802.11. In this case, technologies such as xDSL are regularly used (for example, HDSL) and lines T1, T3, E1, and E3, among others. Technologies and protocols, such as HDLC or PPP, are used at level 2. In this type of connection there is no other type of level 2 equipment or higher (e.g., switches and routers) between the two extreme points of the connection (customer's LAN and carrier). What can indeed exist are different types of layer 1 equipment such as repeaters, amplifiers, and multiplexers. Figure 3.6 shows the equipment traditionally used to carry out the clear-channel connection. In this case, the connection of

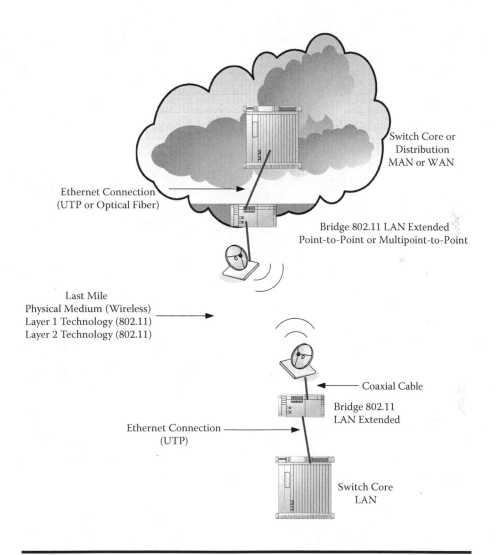

Figure 3.3 LAN extended solution.

the switch to the router can be seen through Ethernet and that of the router to the model through a V.35 interface.

We will now mention some HDLC and PPP basic characteristics. HDLC is a data link protocol geared toward bits that is in the second layer of the OSI model. It has been widely implemented given that it supports half-duplex, full-duplex transmission types in point-to-point and multipoint networks in addition to switching and nonswitching channels.

Different types of stations have been defined in HDLC as follows:

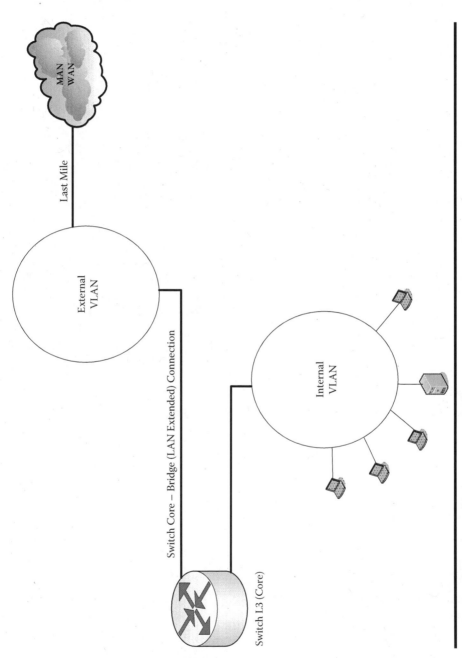

Figure 3.4   LAN extended with broadcast domains.

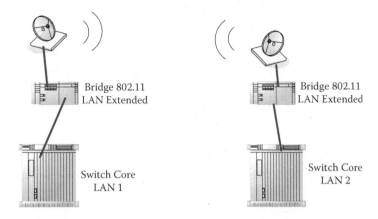

Figure 3.5   Point-to-point solution.

Primary. These are responsible for controlling the other stations (secondary) for the link operation, organizing the data flow control, and detecting errors and recovering.

Secondary. The secondary stations are under the control of the primary ones, are not capable of controlling the link, and are activated only when the primary stations so require.

Combined. This is a combination of primary and secondary stations. Combined stations are capable of sending and receiving commands and responses without the other stations' permission; in other words, they have control over themselves.

The channels should be shaped as follows:

Unbalanced Configuration. The unbalanced configuration in HDLC consists of a primary station and one or more secondary stations. The unbalanced

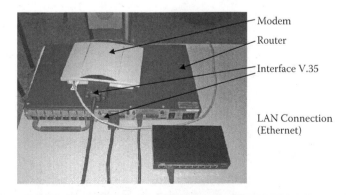

Figure 3.6   Clear-channel equipments.

condition is due to the fact that a station controls the rest. This type of configuration can be used in point-to-point or multipoint networks, in the full-duplex or half-duplex modality.

Balanced Configuration. The balanced configuration in HDLC consists of two or more combined stations, each of which has equal responsibility over the link control. This type of configuration can be used in point-to-point networks or in the full-duplex or half-duplex modality.

Asymmetric Configuration. The asymmetric configuration in HDLC consists of two unbalanced point-to-point independent configurations. Each station has a primary and a secondary state and is considered as two stations.

Finally, the channels can have three operating modes:

Normal Response Mode (NRM). The primary station starts the transmission toward the second station. The secondary station transmits only if it receives a specific permission from the primary station. (The unbalanced configuration is used in this case.)

Asynchronous Balanced Mode (ABM). This uses combined stations. (A balanced configuration is used.)

Asynchronous Response Mode (ARM). This uses a combination of the two previous modes.

An HDLC frame consists of the following fields:

Delimiting fields—a sequence of flag bits that indicates the start and the end of the frame (flag octet: 01111110).

Addresses field—used to identify the station that has transmitted or is receiving the frame (8 bits or more). In PPP this field is 11111111.

Control field—used for sequence numbers and acknowledgement frames, among others.

Data fields—are of variable length and can include any type of date.

Checksum—a variation of the Cyclic Redundancy Code (CRC-CCITT) that allows the detection of bytes of lost indication.

There are different types of frames in HDLC in order to conduct the data transmission, as well as the transmission of control data between transmitters and receivers.

Information frames—transport the user's generated data

Supervision frames—provide acknowledgments, which are used to indicate rejection, receiver that is not ready, receiver that is ready, and selective rejection

Non-numerical frames—provide complementary functions for link control, such as setting a normal response, asynchronous, disconnected balanced asynchronous, non-numerical confirmation, and disconnected and disconnection request mode.

Figure 3.7 shows the connectivity physical scheme through a clear-channel connection. The switch core of the LAN network connects through this type of connection to the router by means of an Ethernet port, traditionally with UTP, except if the distance is greater than the one supported by this type of cable. The router is connected to the modem by means of a V.35 interface, and this physical

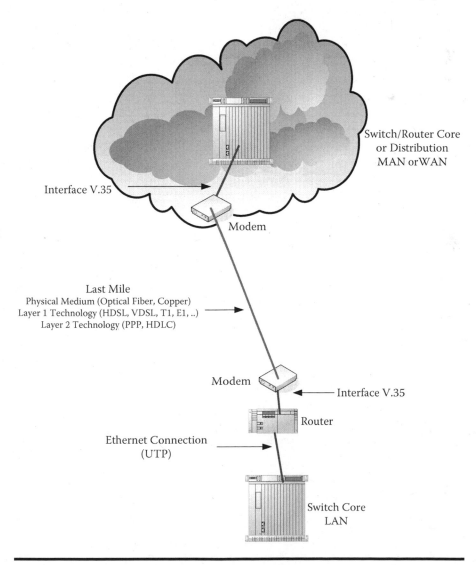

Figure 3.7   Clear-channel solution.

communication takes place between the customer's LAN network and the carrier inlet through the last mile, whether it is through physical means in copper or optic, or through microwave or radio. The inlet into the carrier's site will be done the same way it's done in the customer's LAN network and in a way that is similar to the one explained in the LAN Extended solution. The objective of the carrier's inlet is to connect to the backbone, and usually this connection is made through distribution equipment that is in charge of connecting many access networks or last miles to the core equipment, or in several cases the connection is made directly to the equipment that is part of the carrier's backbone.

If the connection is made through a physical solution such as microwave or radio link, the modem will need to be changed and the corresponding antenna installed as shown in Figure 3.8. In this case, layer 2 will continue being the same; in other words, it could be, for example, HDLC or PPP.

Given that there is a router between the LAN network and the last mile, the problem mentioned in the LAN Extended solution is not seen here given that the router carries out the division function of the broadcast domains. Therefore, the LAN network's broadcast will remain only in this area and will not exit through the clear-channel connection (Figure 3.9). It is clear that different broadcast domains can exist within the LAN network through the VLANs in the switches, but this is a topic already covered in Chapter 2.

Also in LAN Extended, another connection could be the use of a point-to-point communication between two organization sites, in which case a carrier network will not be used. Figure 3.10 shows this scheme.

3.1.3 ADSL

There are many types of carriers that in one way or another carry connectivity and services to our homes and businesses. This is, for example, the case for telephone or cable TV carriers. Data transmission technologies were enhanced so the same copper cables could transmit Mbps and not only Kbps. Through ADSL technology it is possible to transmit different Mbps through the same telephone lines and through the same coaxial cable through which TV channels are transmitted, and it is possible to transmit different Mbps through cable modem technology.

This section analyzes ADSL technology by means of which we can connect in the last mile to the carrier. The cable modem case is very similar, but it will not be covered in this book, even though it is a good last-mile option.

As mentioned in Chapter 1, ADSL is a technology that works using frequency multiplexing by means of which it is possible to make a phone call and at the same time have data transmission to (upstream) and from (downstream) the carrier. This section will explain only the part on data transmission using the ADSL network. Traditionally, ADSL works with two types of equipment: the switch with ADSL ports whose name is *Digital Subscriber Line Access Multiplexer* (DSLAM), which is

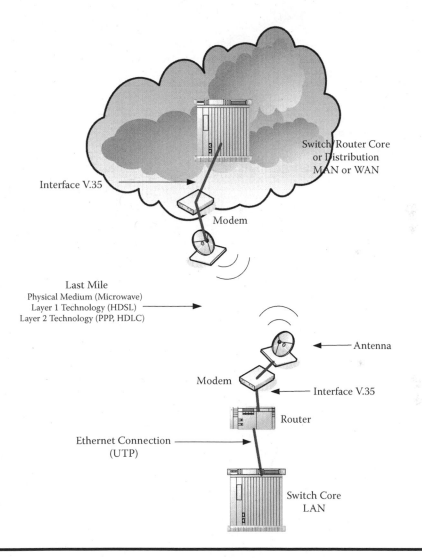

Figure 3.8   Clear channel with wireless solution.

located in the carrier, and the modems or bridge or ADSL routers that are located on the customer's side.

Prior to explaining the network's designs through ADSL we will mention some characteristics of DSL technologies and their evolution.

xDSL is a set of technologies that allows the transmission and receipt of data at high velocity, using analog, conventional telephone line. Conventional telephone lines have a bandwidth of 4 KHz, which is sufficient to transmit voice in base band. Using a conventional modem with this bandwidth and transmission, speeds of 33.6 Kbits/s and, in optimum conditions, even 56 Kbits/s can be achieved.

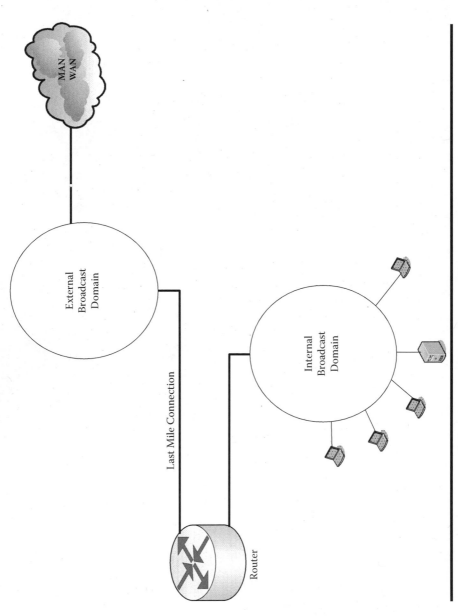

Figure 3.9   Clear channel with broadcast domains.

Figure 3.10 **Point-to-point solution with clear channel.**

In order to increase transmission velocities and to get the maximum potential out of such a large network, as is the case of the telephone network, several xDSL technologies were created that convert conventional analog line into high-speed digital line, using multiplexing and modulation techniques. The telephone network, made up of copper lines, which are in massive use, is for ideal covering data transportation to large populated areas. Conventional telephone modems provide users asymmetrical transmission rates, whereas xDSL technologies can be asymmetrical or not; in other words, they adapt to each user's requirements. However, this type of technology has an inconvenience and that is the limited distance between the subscriber and the company providing the service. This distance cannot be greater than 6 Km and it may affect the service quality. xDSL systems are known for multiplexing the different signals to be transported; a typical xDSL system is made up of three channels: a voice channel, an upstream channel, and a downstream channel. Basically, these three channels are separated by two devices. The voice channel is isolated by means of a splitter, which is responsible for dividing the high frequencies from the low ones, while the modem is responsible for modulation and demodulation of the two data channels (upstream and downstream channels). We have different types of DSL technologies, such as the following:

HDSL (High Data Rate Digital Subscriber Line). This was a technique designed for transmission through copper cables of T1 frames (United States) or E1 (Europe), which is made up of different telephone channels. It withstands flows of up to 2.048 Mbps. The only inconvenience is that it requires several copper pairs: two for the transportation of a T1 frame and one for an E1 frame. Therefore, it cannot be used in the subscriber loop (which has only one pair), and it is mainly used among switching or mobile telephone base stations.

SDSL (Single Digital Subscriber Line). This is an HDSL version that uses only a copper pair. Moreover, it allows the simultaneous use of the plain old telephone system (POTS), in other words, the traditional basic telephony service. Therefore, it can be perfectly used in the subscriber's loop. It is an asymmetrical operation; in other words, the assigned bandwidth is the same in the subscriber-network sense (upstream) as in the network-subscriber sense (downstream).

ADSL (Asymmetrical Digital Subscriber Line). This is similar to SDSL, but in this case the operation is asymmetrical, giving a greater bandwidth to downstream communication (around 1 MHz) than to upstream (around 110 kHz). It works like this because it is a technology focused exclusively on its use in the subscriber's loop and most of the services required by users need such asymmetry: Internet access, video-on-demand, telemarketing, etc. For example, a MPEG video transmission requires up to 3 Mbps in the downstream link and only 64 Kbps in the upstream. The transmission speeds achieved with ADSL could be up to 1 Mbps in the upstream link and 8 Mbps in the downstream.

ADSL2. This is an ADSL evolution that allows a capacity increase of up to 1 Mbps in the upstream link and 10 Mbps in the downstream. This is achieved mainly by the use of greater complexity in the signal processing techniques (with so-called Trellis modulation) and header reduction.

ADSL2+. This keeps the innovations introduced by ADSL2, having as its main novelty the duplication of the bandwidth used, extending it from 1.104 kHz (of ADSL and ADSL2) up to 2.208 kHz. This enables duplicating the downstream link capacity up to 24 Mbps, even though the upstream is kept at 2 Mbps.

VDSL (Very High Data Rate Digital Subscriber Line). While the ADSL technology covers the entire customer's loop, VDSL technology aims at covering only the last meters of the mentioned loop (as a maximum 15 km). This enables a significant increase of binary rate (up to 2 Mbps in the upstream links and 52 Mbps in the downstream). VDSL must go hand-in-hand with the FTTC (fiber to curb) technology, and most of the customer's loop is replaced by optic fiber, which links the telephone station with a device called an *optical network unit* (ONU), the English-speaking term, or *optical network terminal* (ONT), located very close to the customer's house. The pair cable remains there from the ONU to the user, and basically this is where VDSL is used. The ONU carries out the optic-electric conversion and vice versa.

As far as ADSL, we mentioned that it is a bandwidth technology that enables the computer to receive data at a high speed, all this through a conventional telephone line by means of the data signal modulation used by the computer. One of the ADSL characteristics, which contributed to the use of this technology in Internet, has been that it is an asymmetrical system in which the transmission speed in both directions is not the same. In an Internet connection, usually the downstream transmission speed (Internet→Host) tends to be greater than the upstream (Host→Internet). An example of this is access to a Web page. In order to do this we must make a request to the corresponding server indicating we want to access the page, which is done with the transmission of a few bytes, while the server sends us the whole page, which could use from a few Kbytes to many Mbytes. Therefore, it is necessary to count on a higher download speed. The first specification on the xDSL technology was defined by Bell Communications Research, a pioneer company in Integrated Services Digital Network (ISDN), in 1987. At first this technology was developed for the supply of video-on-demand and interactive television applications. The current ADSL was developed in 1989. The arrival of this new technology for communications in Spain took place only six to seven years ago, with the implementation of a full rate through the copper pair that is used for telephone purposes. The government freed the telecommunications market and this led to conflicts because this allowed other companies to provide Internet services based on the ADSL technology, but the main part of this technology, which is the subscriber's loop, still belonged to Telefonica, which by then had a monopoly over communications in Spain, and the latter rented the subscriber's loop to different companies so they could exploit it. Given there were few cable operators then and the technology was not needed by the subscriber's loop, management forced Telefonica to provide a telephone infrastructure, which allowed the exploitation of these services at a high speed. Consequently, with time many companies started offering Internet services under ADSL, which promotes competition and leads to a price reduction.

ADSL is a signal modulation technology that allows high-speed data transmission through a couple of copper threads (telephone connection). The first difference between 56K modem modulation and ADSL is that these modulate at a frequency range greater than the regular ones, 24–1104 kHz for ADSL and 300–3400 Hz for the regular ones, as well as with voice modulation. This assumes that both types of modulation can be active simultaneously given that they work at different frequency rates. The ADSL connection is an asymmetric connection and the modems located in the station and in the user's house are different. Figure 3.11 shows how an ADSL connection works. We see that the modems are different and in addition to this, there is a splitter between them, which is made up of two filters. One is for the upper passage and the other one is for the lower passage, and its only function is to separate the two signals that go through the transmission line, that of voice (vocal) telephony (lower frequencies) and that of data (high frequencies). As seen before, ADSL needs a pair of modems for each user: the one the user has at home and the other one at the operator's station. This duplicity complicated the deployment of this access technology in the local stations where we had the connection of the subscriber's loop. DSLAM was created to solve this problem. The latter consists of a cabinet that includes various ATU-C modems, and it gathers all the ADSL subscribers' traffic toward a WAN network. Thanks to this technology, the deployment of the modems in the stations has been pretty simple, and consequently ADSL has been widely extended.

We will explain below the ADSL connection scheme. The connection with an ADSL takes place as follows. The bridge or router with an ADSL WAN port is connected to the LAN network (core switch) through an Ethernet port. The ADSL router is connected through the ADSL port (RJ-11) to the telephone line. In the event that it is necessary to make telephone calls simultaneously, it is necessary to install a splitter to separate the data connection to the telephone connection in order to transmit voice. The ADSL line takes you to the carrier in the last mile. ATM is used as a layer 2 technology in this last mile to establish the virtual circuit. This ADSL line goes into the distribution or aggregation equipment in the carrier, which is called *DSLAM*. The connection to the backbone is done through this DSLAM, whether it is through a SONET port or SDH, or the connection to the backbone of the MAN or WAN network to the carrier can be made from an Ethernet port. Figure 3.11 shows the connection scheme. It can be seen in this figure that through the DSLAM with the splitter it is possible to receive many connections from the last mile through different ports and ADSL lines, and through different virtual circuits with VPI/VCI values in ATM. These configurations of VPI/VCI values

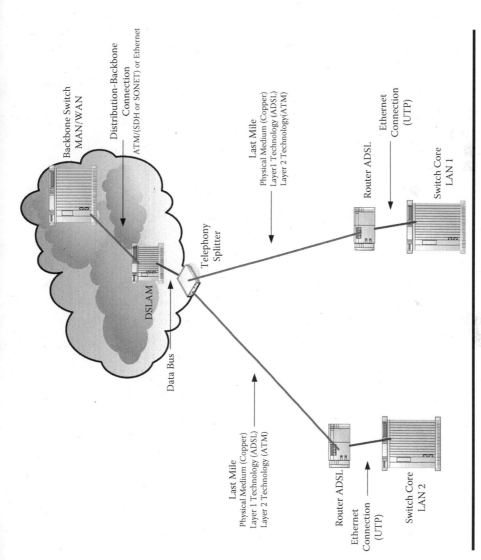

Figure 3.11 ADSL connection.

will be included at the end of this chapter where designs and complete cases of connectivity, even with IP, will be shown.

As previously mentioned, it is necessary to define an ATM virtual circuit with this ADSL technology. Therefore, it is necessary to shape a VPI/VCI value, and on the customer's network side (LAN) it is necessary to define in the ADSL WAN port the same VPI/VCI value defined in DSLAM. Figure 3.12 shows a virtual circuit design for the two connections shown through ADSL. As seen in Figure 3.12, two different virtual circuits have been defined to carry out the two ADSL connections. The virtual circuit that connects LAN 1 with the carrier has the VPA/VCI 0/100 value and the virtual circuit that connects LAN 2 with the carrier has the value of VPI/VCI 1/200.

Figure 3.13 shows the back of the DSLAM ADSL, which shows the inlet of the ADSL lines to the splitter and the data bus connection to the DSLAM. In this case the splitter has RJ-11 connections.

Figure 3.14 shows the front end of the DSLAM, which shows the management and uplink ports in order to carry out the connection toward the MAN or WAN backbone. In this case the uplink is done through Ethernet. (The figure shows that this could be in optical fiber or in UTP.) This connection could be made, as previously mentioned, by means of an SDH or SONET port.

Figure 3.15 shows an ADSL bridge or router, which would be located in the customer's LAN network as an exit (outlet) toward the last mile. In this case the ADSL equipment has an Ethernet port to connect to the LAN and an ADSL port with an RJ-11 to connect to the last mile.

3.1.4 Frame Relay

As far as Frame Relay, we will start by mentioning certain characteristics of this technology and we will explain later how the connectivity scheme and design operate through Frame Relay.

Frame Relay is a low-performance WAN protocol when compared to the old ATM and the current MPLS. It operates in the Data-Link layer of the OSI reference models. Initially, Frame Relay technology was designed to be used through the Integrated Services Digital Network (ISDN). Today it is also used through a wide variety of other network interfaces.

Frame Relay is an example of packet-switching technology. The networks using this technology have terminal stations that share the network's transmission means dynamically, as well as the bandwidth available. The packets of variable length are used for more efficient and flexible transfers. Further on, these packets are switched among the different network's segments until they get to their destination. The statistical multiplexing techniques control the access to the network in a package-switching network. The advantage of this technique is a more flexible and efficient use of the bandwidth. A Frame Relay is often described as an X.25 compact version with fewer characteristics as far as robustness, such as sliding window management

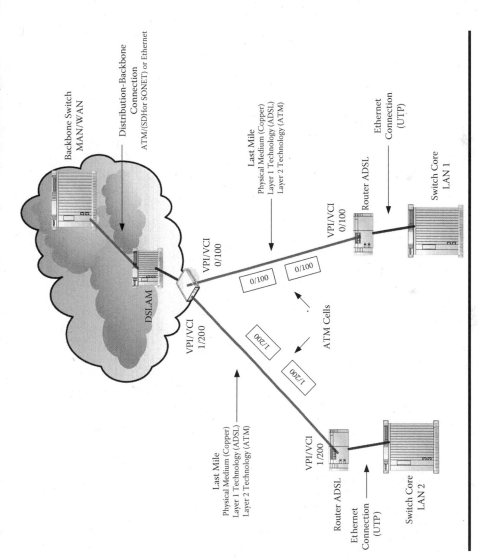

Figure 3.12   ADSL with virtual channels.

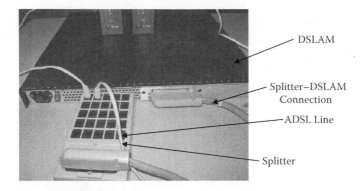

Figure 3.13   ADSL DSLAM (back).

Figure 3.14   ADSL DSLAM (front).

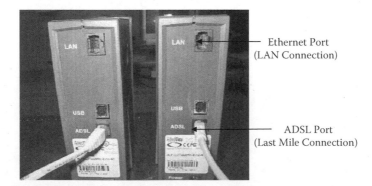

Figure 3.15   ADSL bridge/router.

and the more recent data retransmission. This is due to the fact that Frame Relay usually operates through WAN installations that offer more reliable connection services and a higher degree of reliability than those available in the 1980s, which were used as usual platforms for the WAN X.25. As previously mentioned, Frame Relay is strictly a layer 2 architecture, whereas X.25 also provides layer 3 services (the network layer).

Frame Relay is a data packet transmission protocol in which data packets are transmitted in high-speed bursts through a fragmented digital network in transmission units called *frame*. Frame Relay requires an exclusive connection during the transmission period. This is not valid for video and audio transmissions given that they require a constant transmission flow. Frame Relay is a rapid packet technology given that the error checkup doesn't happen in any transmission node. The extremes are the ones responsible for checking errors, but errors in digital networks are less frequent when compared to analog networks. A rapid packet is transferred in the Asynchronous Transfer Mode (ATM) with each Frame Relay or with a transmission element. Frame Relay transmits packets at the data forward level of the open system of interconnection (OSI) before at the network level. No matter if the packet is of fixed size, a frame is variable in size and can be as long as 1000 bytes or more.

Almost all Frame Relay users are using it for the same reason: the LAN's interconnection. It is not surprising that Frame Relay was designed for that same reason. Prior to Frame Relay, in most cases rented lines or fixed channels in controlled bandwidth networks were used. Some inter-LAN traffic is transported in X.25; however, this tends to be specific for low-bandwidth applications. Many X.25 networks were designed some years ago to support low-bandwidth applications and therefore they are less convenient for large-bandwidth and critical time applications, which are usually utilized in LANs. This is therefore an alternative for rented lines of a fixed bandwidth. In spite of this, Frame Relay is not only used for cost savings, but also for better service quality. A Frame Relay can be highly viable because it offers the possibility of choosing a new route in case of line failure, and consequently, because it has a rich interconnection pattern, Frame Relay can reduce the number of hops between intermediate nodes giving unforeseeable response times. Frame Relay frames and headers may differ in lengths, given that it offers a wide variety of available options in the implementation, known as annexes to the definitions of the basic standard. The information transmitted in a Frame Relay frame may fluctuate between 1 and 8.250 bytes, but because of a flaw it does it at 1.600 bytes. Frame Relay networks are geared toward the connection, as are X.25, SNA, and even ATM. The connection identifier is the union of two fields of two HDLC fields, on which Frame Relay is based for original specifications of data units (layer 2 protocol). Some control bits (CR and EA) are inserted between the two HDLC fields that form the *data link connection identifier* (DLCI). Then other fields with very special functions in the Frame Relay networks are added. This is due to the fact that the Frame Relay switch nodes lack a packet structure in layer 3, which is generally used for the implementation of functions, such as flow control and network congestion, and these functions are necessary for the proper operation of any network. The three most essential ones are Discard Eligibility (DE), Forward Explicit Congestion Notification (FECN), and Backward Explicit Congestion Notification (BECN). The DE bit is used to identify frames that could be rejected in the network in case of congestion. FECN is used with end system protocols that control the data flow between the transmitter and the receiver, as the windowing mechanism or TEC/IP; in theory, the receiver may adjust its "window" size in response to the frames arriving with the activated FECN bit. Obviously, BECN may be used with protocols that control data flow from one extreme to the other in the transmitter itself. According to this, the network is capable of detecting errors, but not of correcting them, and in some cases it may even eliminate frames.

For a long time, while the Frame Relay services were receiving the interest of European users in specific applications, such as the interconnection of local networks, the operators saw that it is necessary to discourage its use to support voice communications. For technological and regulatory reasons, those were the rules of the game to which the users have been adhering. Consequently, a new Frame Relay supply came about that tried to exploit the indubitable appeal that assumes supporting in the same line the company's voice, fax, and data transmissions with the consequent economic (better use of the bandwidth, full rate, etc.) and control (a single operator and service manager) advantages. And if circumstances advise it, there is always the possibility for the organizations to install, operate, and manage by themselves their own Frame Relay networks.

Many users found an inexpensive solution to transport their data with the migration of point-to-point networks to Frame Relay. Nevertheless, since many of these dedicated lines were also used to transport voice, it was necessary to look for a solution to the corporate voice needs.

Voice over Frame Relay (VoFR) allows those responsible for communications and those in charge of the network to consolidate voice and data, such as fax and analog modems, with data services in the Frame Relay network. The end result is a potential cost savings and the simplification of the corporate network management and operation. VoFR manufacturers and users have stated that in addition to the savings, VoFR offers a quality comparable to the one obtained in traditional voice networks (toll quality).

The Implementation Agreement (IA) was ratified by the Frame Relay Forum. FRF.11 indicates how voice is transported in a Frame Relay network and enables salespeople to develop services and equipment that can interoperate. This IA allows for those responsible for a network that are looking to reduce their communication infrastructure costs and maximize their return on investment of their Frame Relay network to consider VoFR as an alternative to traditional voice services.

In some cases, the users could discover they have excess bandwidth in their Frame Relay framework that could efficiently support voice traffic. Other users could discover that, with a cost increase to acquire additional bandwidth in their

Frame Relay connections to include voice traffic, they are actually saving money vis à vis traditional telephone services offered by local or long-distance vendors.

VoFR may offer an economic solution to the transportation of voice traffic among the different company's offices. For example, the person responsible for the network may integrate some series voice and data channels in the same Frame Relay connection between the company's remote offices and the central office. By combining data and voice traffic in an existing Frame Relay connection, the user has the possibility of making economic phone calls within the company and giving an efficient use to the network's bandwidth.

Another example of applications in real time on Frame Relay is video, which is becoming more popular with time, fostered by the introduction of video coders with an integrated Frame Relay interface. This is critical because videos require a constant data flow so they can be delivered without a variation in the delay (jitter), without shadows and image distortions, in order to obtain a good quality video, given that the coders manage and store the monitor's image, and do compression and make changes to the image from video segments to segments.

The world is seeing a growth in the interest of video applications over Frame Relay given that users are weighing the economic advantage and the Frame Relay availability with dedicated lines or ISDN, mainly in those parts of the world where ISDN is not available. In order to give a solution to video, it was specified in the ATM networks. The integration of voice, fax, video, and data traffic in a single access line offers a viable and interesting option for those responsible for corporate networks.

We have made some comments on Frame Relay technology; now we will specify the connection and design scheme.

This layer 2 technology, which works by means of a virtual circuit configuration, enables making connections to different sites, for example, from a main site through one physical connection. Likewise, this type of connection can be made through ADSL with ATM. Once the core is analyzed in this section and a complete solution that connects different sites through the last miles is shown, we can see the effect of shaping different virtual circuits. This section covers last-mile connectivity only as part of this complete scheme.

A Frame Relay connection works as follows: The router with the WAN port, usually with serial ports, ISDN, and E1, among other layer 1 technologies, is connected to the LAN network (core switch) through an Ethernet port. The router connects through the serial port (V.35), ISDN, lines E1, etc., to the base band modem, which connects to the physical means established in the last mile by means of the carrier. Layer 1 technologies, such as HDSL and lines E1 and T1, are used in this last mile, as we have previously mentioned. Frame Relay is used in this last mile as a layer 2 technology for the establishment of the virtual circuit. This virtual circuit goes into the distribution or aggregation equipment (which is at least a switch, but may also have the routing function) of the carrier, which is usually a multi-platform equipment given it has different level 1 and level 2 connection platforms. The connection to the backbone of the carrier's MAN or WAN network is done through this switch, whether by means of a SONET or SDH port or an Ethernet port. Figure 3.16 shows the connection scheme. As can be seen in this figure, it is possible to receive many last-mile connections through the switch located in the carrier, and this is done through different ports and physical lines.

Through each last mile's physical connection it is possible to make different logical connections through the virtual circuit. Each virtual circuit is specified through the DLCI values in Frame Relay. As can be seen in Figure 3.17, the virtual circuits leave the LAN 1 network toward the carrier, the first one with a DLCI 20 value and the second one with a DLCI 22 value. This shows how, through the Frame Relay technology, which carries out the switching function, it is possible to make several logical connections through only one physical connection. These end-to-end configurations of DLCI values can be seen in the last part of this chapter, which covers designs and complete connectivity cases, even with IP.

Figure 3.18 shows the equipment traditionally used to carry out the connection by means of Frame Relay layer 2 on the customer's side in its connection to the last mile. In this case, the switch connection to the router can be seen through Ethernet and from the router to the modem through a V.35 interface. These modems are mainly HDSL in layer 1 using copper pairs. As can be seen in this type of connection, this example is the same as the clear-channel connection; the difference at the equipment lies in the fact that this last-minute connection in the carrier will be received by an equipment that will switch through Frame Relay.

Figure 3.19 shows some equipment as examples that will be part of the Frame Relay inlet in the carrier. Here we can see the Frame Relay switch receiving connections through some Winchester connectors and through the splitter making the connection to the switch.

3.1.5 WiMAX

As with the previous technologies, we will first start with some basic comments about WiMAX technology and then we will explain the design and connectivity scheme.

WiMAX (Worldwide Interoperability for Microwave Access) is the name given to the 802.16 standard that describes the Air Interface for Fixed Systems of Broadband Wireless Access. The 802.16d standard, already homologated, corresponds to the fixed wireless transmission. The 802.16d standard is a variation of the fixed standard with the advantage of optimizing the power consumption and thus reducing the customer premise equipment (CPE) modem size. We then found the 802.16e, which is responsible for mobile wireless transmission.

WiMAX was designed as a last-mile solution in MANs in order to provide services at the commercial level. It can provide every level of services necessary for a carrier, depending on the agreement with the subscriber, different packet services such as IP and VoIP, and switching services such as TDM, E1s/T1s, traditional voice, ATM interconnections, and Frame Relay.

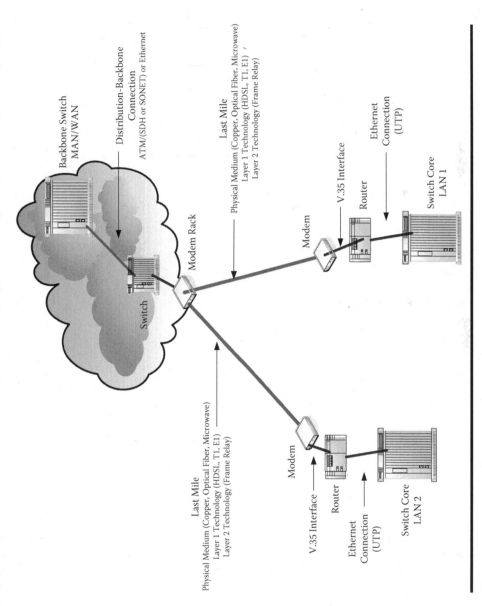

Figure 3.16 Frame Relay connection.

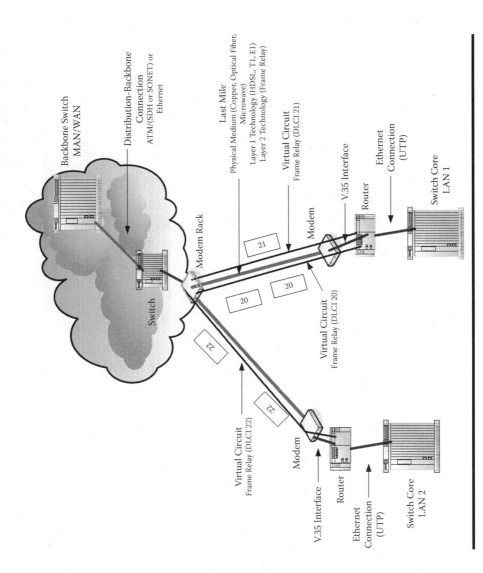

Figure 3.17 Frame Relay with virtual channels.

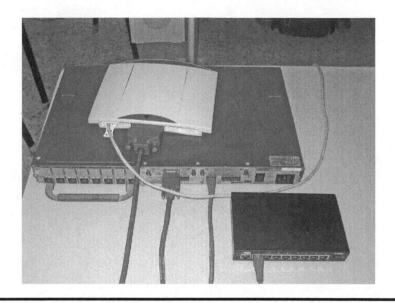

Figure 3.18   Frame Relay equipments (client connection).

Given that the 6–11 GHz frequency range requires a line of sight, its use is planned for transmission among antennae, but the 2–6 GHz range could be used for direct distribution to end users. Transmission is more vulnerable to climate conditions at high frequencies, but higher speeds are obtained. The use of low frequencies (less than 700 MHz) is analyzed even if the transmission speed experiences a reduction because of the use of error correction protocols. At the end of the day, the actual speed will depend on the number of users simultaneously connected and on broadband requirements of each of them.

The maximum WiMAX coverage radio (30 to 50 km) will depend on the presence of buildings, mountains, and large obstructions in the middle. However, this

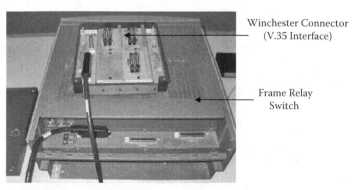

Winchester Connector
(V.35 Interface)

Frame Relay
Switch

Figure 3.19   Frame Relay equipments (carrier).

is already progress because that coverage is not achieved with any other wireless technology.

The bright future of this wireless technology responds to the support it has received from over 130 organizations: equipment and components manufacturers, service vendors, software programmers, etc. Intel Corporation manufactures chips that work as brains in over 80% of the world's computers; because of this reason its dissemination has been very significant. Other companies, such as Nokia and Siemens, are already placing chipsets in their future versions of mobile telephony and PDAs.

Nowadays, the expected technology to be massively launched is the standard WiMAX Mobile. The 802.16e standard of the IEEE is a reviewed version of the 802.16-2004 base specification, which is geared toward the mobile market, adding portability and capacity to mobile customers with IEEE. The 802.16e standard uses Orthogonal Frequency Division Multiple Access (OFDMA), which is similar to Orthogonal Frequency Division Multiplexing (OFDM), given that it divides the multiple subcarriers.

However, OFDMA goes one step beyond gathering multiple subcarriers in sub-channels. A single subscriber's customer station may use all the subchannels within the transmission period, or multiple customers could transmit simultaneously, each one using one portion of the total number of subchannels. The 802.16-2004 standard of the IEEE enhances the last mile forward in different crucial aspects: multipath interference, disseminated delay, and robustness. Multipath interference and delay enhance the actions in those situations where there is no direct sight line between the base station and the subscriber's station.

The emerging media access control of 802.16-2005 is optimized for long-distance links because it has been designed to withstand longer delays and delay variations. The 802.16 specifications adjust media access control management, which enables the base station to question the subscribers, but there is a certain amount of delayed time. WiMAX equipment managed in the frequency bands, which are exempt in the license, uses time division duplication (TDD). The equipment operating within the authorized frequency bands will use TDD or frequency division duplication (FDD). The 802.16-2005 of IEEE uses an OFDMA for the optimization of data wireless services. For this case the OFDMA signal is divided in 2048. As previously indicated, the more subcarriers on the same band, the tighter the subcarriers are.

Table 3.1 shows a summary of the general aspects found in the .16e version of the IEEE 802.16 standard.

802.16e incorporates new signal's power adjustment techniques and information coding that enhance yield. This enables mobility and the use of equipment in interiors. 802.16e is by far the best current alternative for high-capacity Internet access by the ISPs. It can act as a WiFi supplement or it can fully replace it.

As far as its architecture, it shows a direct correlation with its predecessor 802.16-2004, where links with lines of sight can be established to interconnect the base

Table 3.1

Characteristics	802.16e
Spectrum	< 6 GHz
Functioning	Without direct vision (NLOS)
Bit rate	Up to 5 Mbit/s with 5 MHz channels
Modulation	OFDMA with 2048 QPSK, 16QAM, 64QAM subcarriers
Mobility	Vehicle mobility (up to 120 Km/h)
Bandwidth	Selected between 1.25 and 20 MHz
Typical cell radio	Around 2–5 km

station (BS) of WiMAX Mobile. Moreover, it can provide services included within the IP applications for mobile workstations (PDAs, portables, vehicle information systems, among others), and in addition to this an interconnection with existing WiFi technologies can be achieved. Intel is currently working on marketing the first phase of the chipset for devices that are compatible with the 802.16e standard, the PCMCIA cards are an example of this.

Now that we have made some comments on WiMAX technology, we will proceed with comments on connection and design scheme.

As known, WiMAX is a widely used wireless solution for connectivity in fixed connections and, in the short term, in mobile connections in the metropolitan scope. From the fixed standpoint, WiMAX is a good solution that may be used in the last mile. This means and implies that there is a carrier, whose last mile solution is under WiMAX coverage and the connections of base stations are carried out through high-speed metropolitan rings that, as previously mentioned, could be ATM/SONET or ATM/SDH or Metro Ethernet, taking into account, however, that ATM/SONET or ATM/SDH is a solution on the decline.

We have a bridge or router available to make this connection from the customer, whose exit to the last mile is done through WiMAX. This bridge or router connects to the main switch of the LAN network. On one hand, the WiMAX bridge or router connects wireless, through the antenna, to its corresponding base station according to the coverage ratio of each base station. These base stations will be connected further on, which could be done through Ethernet to the core of the carrier's network. Figure 3.20 shows a connection scheme through WiMAX.

3.1.6 Ethernet Access

We are presenting in this section the scheme when the last mile is connected through Ethernet, which is being widely used given it is easier to do so. In this

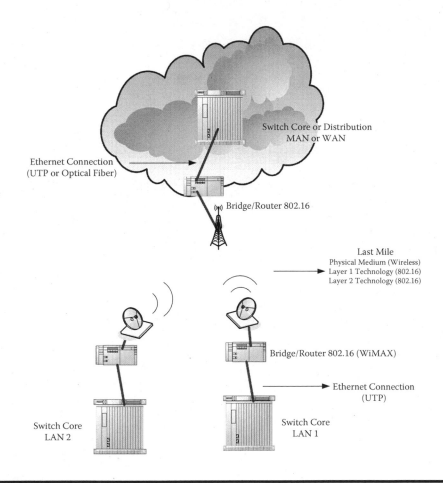

Figure 3.20 WiMAX connection.

case, the connection will be made taking one of the ports of the main switch and through the optic fiber making the connection to the carrier. Figure 3.21 shows the scheme for this type of connection.

3.2 MAN/WAN Core Solution

We have already seen several forms of last-mile connection to the carrier; now we will analyze the carrier's backbone connection scheme. Traditionally, this backbone has been designed under a transportation technology in SONET or SDH and the switching is done through ATM. Other alternatives arose with time, such as POS (Packet over SONET); at the present time many carriers worldwide have implemented Metro Ethernet as another solution to the backbone of networks and

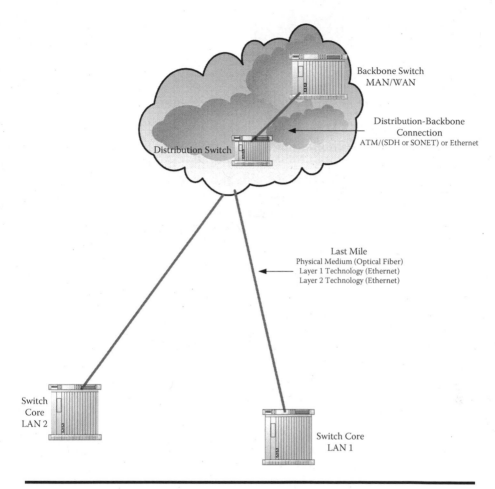

Figure 3.21   Ethernet connection.

the best thing is that, as time elapses, Metro Ethernet has more followers. Even if that's the case, SONET or SDH are not disappearing, but what will indeed disappear as a switching technology is ATM; actually, what we have seen is that both technologies SONET or SDH and Ethernet have been adapted to implement DWDM and through this the carriers' transmission capacity has increased. As part of this thematic, it is worth mentioning Multiprotocol Label Switching (MPLS) as a technology by means of which we can switch, but applying the quality of service (QoS) concepts in a more straightforward and direct way between IP and level 2, unlike those cases in which this technology doesn't exist in the carrier. The optical switching can be seen as the future in these carriers' backbone networks through wavelengths, and in order to carry out the implementation and functionality of this technology different switching schemes have been proposed: Optical Packet Switching (OPS), Optical Lambda Switching (OLS), Optical Circuit Switching

(OCS), and Optical Burst Switching (OBS), and it is precisely in this subject where Generalized Multiprotocol Label Switching (GMPLS) plays an important role as a technology that defines the control plan to achieve this switching.

3.2.1 ATM (SONET/SDH)

We will now analyze ATM working together with Synchronous Optical Network (SONET) or with Synchronous Digital Hierarchy (SDH) as the first technology related to the carrier's backbone in MAN or WAN networks. This is the typical voice carrier's network that was converted as a result of the convergence process for data transmission.

As done with previous technologies, first we will make some basic comments on SONET or SDH and ATM technologies and then we will explain the design and connectivity scheme.

SONET is a standard for transportation via optic fiber that was formulated in 1984 by the Exchange Carriers Standards Association (ECSA) for the American Standards Institute (ANSI), which regulated communication and other industries' standards. This standard was later included in the SDH of the Consultative Committee on International Telegraph and Telephone (CCITT), currently called the International Telecommunication Union (ITU), which regulates international telecommunication standards.

The first generations of optic fiber systems, before SONET, in the public telephone network used their own architectures, equipment, codes, multiplexing formats, and maintenance procedures (each company had its own standard). This equipment's users (Bell regional companies) and companies that interchange carriers (IXCs) in the United States, Canada, Korea, Taiwan, and Hong Kong need a standard that enables them to mix the equipment from different vendors. This task of creating such a standard became the responsibility of ECSA so as to establish a connection standard and connect one optic fiber system with another one. This standard was called SONET.

3.2.1.1 Digital Signal Synchronization

In order to understand the concept and details of SONET, the meanings of *synchronous*, *asynchronous*, and *plesynchronous* must be well understood. Digital transitions in the signal take place exactly at the same time in the set of synchronous signals. However, there is a phase difference between the transition of two signals and this takes place within established limits. This happens because of propagation time and jitter delays introduced in the transmission network. In a synchronous network, all the clocks can be mapped into a primary reference clock (PRC). In the event of two plesynchronous signals, their signals take place at the same time only with a small variation. For example, if two networks must work together, their clocks may come from two different PRCs. Even though these clocks are extremely accurate,

Table 3.2

SONET	Bit Rate	SDH
STS-1/OC-1	51.84 Mbps	—
STS-3/OC-3	155.52 Mbps	STM-1
STS-12/OC-12	622.08 Mbps	STM-4
STS-24/OC-24	1244.16 Mbps	—
STS-48/0C-48	2488.32 Mbps	STM-16
STS-192/OC-192	9953.28 Mbps	STM-64

there is a difference between one clock and the other. This difference is known as *plesynchronous difference*. If two signals are asynchronous, their transitions do not take place at the same time.

3.2.1.2 Basic SONET Signal

SONET defines a technology to carry different signals of different capacities through a flexible synchronous hierarchy. This is possible by means of a multiplexing scheme in bytes (byte-interleaved multiplexing scheme). Such a scheme simplifies multiplexing and offers an end-to-end network. The first step in the SONET multiplexing process involves the lowest level generation or base signal. In SONET, this signal is referred to as synchronous transport signal–level 1 or just STS-1, which works at 51.84 Mbps. Signals at a higher level are whole multiples of STS-1. An STS-*n* signal is made up of *n* byte-interleaved STS-1 signals. Table 3.2 also includes the optical counterpart of each STS-*n* signal, which is called Optical Carrier level *n* (OC-*n*), and the SDH.

3.2.1.3 SONET Characteristics

- Standard for integrated network, based on the optic fiber technology.
- Compatibility in the equipments built for fiber optic management.
- Combines, consolidates, and segregates traffic from different sites through a single installation (preparation or putting a finish).
- In the synchronous aspect of SONET it refers to more stable operations.
- Advanced management of operations, administration, maintenance, and provisioning (OAMP).
- Eliminates the overload of the point-to-point of the ion multiplex (back to back) using new techniques in the preparation process. These techniques are implemented in a new type of equipment called *Add Drop Multiplexer* (ADM).

- Standards for rings management.
- Management of new services transportation such as Asynchronous Transfer Mode (ATM).

3.2.1.3.1 Frame Format

SONET uses a basic STS-1 transmission frame of 51.84 Mbps. Higher-level signals are whole multiples of this base frame. The frame may be divided into two parts: transportation header and synchronous payload envelope (SPE). The latter is then divided into the route header (POH) and the data to be transmitted.

- STS-1 Frame Structure. It can be seen as a matrix of 90 columns × 9 rows. Its first three columns (27 bytes) are for the transportation headers; of these, 9 bytes are for the section header and 18 bytes are for the line header. The remaining 87 columns make up the SPE.
- SPE Structure. It is seen as a matrix of 87 columns × 9 rows. The first column (9 bytes) is used for the POH. Two columns (30 and 59) are not used for payload and actually have no information. The remaining 84 columns (756 bytes) are used for payload.

3.2.1.3.2 Headers and Control Fields

This is information included in the frame to enable the operation of the network and the establishment of an OAMP communication between a controller and individual nodes. The information of the STS-1 headers has several layers:

1. Section Headers. These are used for the communication among elements of adjacent networks (for example, regenerators). They are used for control and signaling purposes. The purpose of these octets is to identify each STS-1 frame start.
2. Line Header. This is seen in STS-n signals among multiplexers. It enables locating the SPE in the frame, multiplexing or uniting two fields, switching automatic protection, and line maintenance.
3. Route Header. This is a header generated in multiplexing points and is managed solely by the initial and final multiplexer. This header manages the state of the route transmission and SPE monitoring.

3.2.1.4 SONET Layers

SONET is based on layers architecture, corresponding to the physical layer of the OSI reference model:

1. Path—offers transportation services among terminal equipments. Its main purpose is to map the services required in the format used by the line layer.
2. Line—main function is to transport the information from the upper layer among ADM. It performs the function of data multiplexing and demultiplexing.
3. Section—responsible for the SONET frames construction and for its transportation through a single optic fiber link. A line is a sequence of one or more sections (united by regenerators) through which the structure of the transported signal in the frames is maintained.
4. Photonics—responsible for the transportation of bits through a physical means and for converting electrical signals into optic signals. Specifies the type of optic fiber and the characteristics of transmitters and receivers.

3.2.1.5 Signals Hierarchy

SONET defines the Optical Carrier (OC) levels, or optic signals, which are the equivalent levels of the Synchronous Transport Signals (STS) or electrical signals for the transportation under optic fiber hierarchies.

SONET supports a concept called *virtual tributary* (VT). VTs are used to support sub-STS-1 levels, which are low-speed signals.

Eventually all the signals are converted into a base format signal (STS-1—51.84 Mbps). This is done in such a way that lower speed signals are the first ones to be converted into VT and then they are converted into an STS-1. After converting them into an STS-1, many of these are multiplexed, in various or in individual STS-n electrical signals.

No other process is required after the multiplexing for processing the signal; the only thing needed is a new conversion from the electrical to optical signal.

3.2.1.6 Physical Elements of SONET

A SONET network is made up of the following:

1. Terminal Multiplexer. This is a terminal equipment used to multiplex other types of digital signals on a SONET network. It is responsible for concentrating any type of lower load signals than the STS-1 on STS-n frames.
2. Repeater. This device is used to regenerate the attenuated signals that travel long distances through the fiber.
3. Multiplexer. Called ADMs, these are intermediate devices that allow multiplexing/demultiplexing digital signals that travel through the fiber without the need of disturbing the other signals. Given that SONET is based on a synchronous technology, it is possible for these devices to know when to enter or exit the signals of its interest from the STS-n frames that travel through the SONET. These devices have a large number of uses: as an intermediate

device to consolidate signals from two different sites, to implement ring-type networks, etc.

4. Digital Cross-Connects. These act as ADMs but they allow more complex connections (with more than two ports). These devices are capable of interconnecting a large number of signals. They work as STS-1 signal switches and therefore are used for consolidation and segregation within the network. These devices are widely used for the implementation of hub-type topologies.

3.2.1.7 Network Topologies

1. Point-to-Point. This is the simplest topology of them all and it enables the easy interconnection of two rapid points, with the disadvantage that it doesn't have the growth possibility offered by topologies such as hub. A SONET point-to-point connection is assembled through the use of the two terminating equipments (Path Terminating Equipment [PTE], such as the terminal multiplexer) united through a fiber trunk, and if necessary regenerators are assembled in case the distances are too long, as shown in the previous figure.

2. Point-to-Multipoint. This configuration allows the distribution of signals that are generated in one point toward consumers located in different sites. It could be used for television broadcast distributions. This type of topology is assembled through the use of ADM equipments, which enable taking the signals needed by the intermediate points without disturbing them so as to continue with the distribution to the following points.

3. Hub Type. This configuration is made up of a central switch that enables creating new circuits among different points. This type of configuration is the more flexible one given that it allows the easy growth of the network. This type of topology is assembled through the use of a digital cross-connect switch (DCS), where the different multiplexers arrive from the different points needed by the network's circuits.

4. Ring Type. This type of configuration is the most widely used for the interconnection of points enclosed in the same geographic area, such as buildings inside a college campus. ADM equipments are needed in order to achieve this type of configuration.

3.2.1.8 SONET Benefits

Among the SONET benefits we can list the following:

1. Network Simplification. One of the best SDH hierarchy benefits is network simplification with respect to networks exclusively based on PDH (Plesiochronous Digital Hierarchy). An SDH multiplexer may incorporate basic traffics (2 Mbps in SDH) at any hierarchy level, without the need of using multiplexers cascade, thus reducing equipment needs.

2. Convergence with Its Modular Architecture. No matter the service, SONET provides broad capacities in terms of flexibility in the service. Many of the broadband services use ATM. ATM multiplexes the load in cells that can be routed as needed. SONET is a logical ATM carrier.

3. Reliability. In an SDH network elements are monitored from one extreme to the other and the maintenance of its integrity is managed.

4. Control Software. The inclusion of control channels within an SDH frame makes possible a total software control of the network. The network's management systems include not only typical functionalities such as alarms management, but other more advanced ones such as performance monitoring, configuration management, resource management, network security, inventory management, network planning, and design.

5. Standardization. SDH standards allow the interconnection of equipments from different manufacturers in the same link.

6. OAMP. This is one of the main tasks of a network supplier. As a result of the continuous growth of networks and the wide variety of manufacturers and types of equipment, there has been a need for centralized management. OAMP is a set of standards that satisfy this need. SONET improves the network's management by providing additional functionality for OAMP in its packet's transmission overhead to simulate communication channels between the networks' controllers or monitors and their nodes and the communication among them. These communication channels are also called OAMP data channels. SONET enables the dynamic configuration of the network from remote sites, that is, to habilitate and rehabilitate, from a remote connection, network's circuits to isolate, block, or redirect traffic due to vendors' problems, petitions, or service policies.

3.2.1.9 SONET Standards

See Table 3.3.

3.2.1.10 Synchronous Digital Hierarchy (SDH)

SDH and its North American equivalent, SONET, are prevailing technologies in the physical transportation layer of current broadband optic fiber networks. Its objective is the transportation and management of many different types of traffic on the physical infrastructure.

SDH is mainly a transportation protocol (first layer in the OSI model) based on the presence of a common temporary reference (primary clock) that multiplexes different signals within a flexible common hierarchy and efficiently manages its transmission through optic fiber, with internal protection mechanisms.

Using OSI as a reference model, SDH is usually seen as a level 1 protocol, in other words, a protocol of the transportation physical layer. In this role, it acts as

Table 3.3

Standard	Description
ANSI T1.105:SONET	Basic description including multichannel, rates, and formats structure
ANSI T1.105.01:SONET	Switching automatic protection
ANSI T1.105.02:SONET	Mapping of useful load
ANSI T1.105.03:SONET	In network interfaces
ANSI T1.105.04:SONET	Data communication channel protocols and infrastructure
ANSI T1.105.05:SONET	Maintenance of connections in cascade
ANSI T1.105.06:SONET	Physical layer specifications
ANSI T1.105.07:SONET	Specification for formats and sub-STS interface rates
ANSI T1.105.09:SONET	Network's synchronization elements
ANSI T1.119:SONET	Communications—OAMP

a physical carrier of level 2 to 4 applications, in other words, the path used for the transportation of higher traffic levels, such as IP or ATM. In simple terms, we can consider SDH transmissions as pipes that carry traffic in the form of data packets.

These packets are of PDH, ATM, or IP applications.

SDH allows the transportation of many types of traffic, such as voice, video, multimedia, and data packets, like those generated by IP. Its role is mainly the same for this purpose: to manage the use of the optic fiber infrastructure, that is, to manage the broadband efficiently while it carries different types of traffic, detects failures, and recovers the transmission in a transparent manner for higher layers.

The main characteristics found today in any implemented SDH transportation network system are the following:

- Digital Multiplexing. This term was introduced 20 years ago and it allowed for analog communication signals to be carried digitally on the network. Digital traffic can be carried much more efficiently and it enables monitoring errors for quality purposes.
- Optical Fiber. This is the most common physical means deployed in current transportation networks. It has a greater capacity for carrying traffic than coaxial or copper pairs, which leads to a decrease in traffic-transportation-related costs.

■ Protection Schemes. These have been standardized to ensure traffic availability. In the event of a fiber's failure or fracture, the traffic could be switched to an alternative route, so the end user will not experience any type of service disruption.

■ Ring Topologies. These are being deployed in larger numbers today. The reason is that if a link is lost, there is always an alternative traffic path through the other side of the ring. The operators may minimize the number of links and of optic fiber deployed in the network. This is extremely important given that the cost of installing new optic fiber cables on site is extremely high.

■ Network Management. Managing these networks from a sole remote site is an important service for operators. Software that enables managing all the traffic paths and nodes from a single computer has been developed. Nowadays an operator can manage a wide variety of functions, such as the supply of capacity in response to customers' demand and monitoring network's quality.

■ Synchronization. Network operators must provide synchronized temporization to all the network's elements to make sure that the information going from one node to the other is not lost. Synchronization is a growing consensus among operators with technological progresses that are more sensitive to time. Synchronization is becoming a critical point, giving SDH an ideal network philosophy path.

3.2.1.10.1 Source

Synchronous transmission systems have been developed in such a way that operators can deploy flexible and resistant networks. Channel insertion and extraction can be made with a single multiplexer. The standard defines the supply of management capacity. Actually, an important agreement has been reached in SDH development. The opportunity for defining this set of standards has been used to address a great number of problems, for example, the need for defining standard interfaces among equipment from different manufacturers and the need for facilitating networks interconnection among transmission hierarchies of North America and Europe.

This standard was completed in 1989 in the ITU-T g.707, g.708, and g.709 recommendations, which define the SDH. In North America, ANSI published its SONET standard, which is known in the rest of the world as SDH standard.

The UIT-T recommendations define a number of basic transmission rates that can be used in SDH. The first of these rates is 155.52 Mbps, usually referred to as *STM-1*. Higher transmission rates such as STM-4, STM-16, and STM-64 (622.08 Mbps, 2488.32 Mbps, and 9953.28 Mbps, respectively) are also being defined.

These recommendations also define a multiplexing structure in which an STM-1 signal can carry a number of lower rate transmission signals forming part of its useful load. Existing PDH signals can be carried on the SDH network as useful load.

The new synchronous standard offered a wide number of advantages, which made it an optimum standard with respect to the previous plesynchronous standard:

- Simple and flexible multiplexing and demultiplexing operations, allowing the extraction and insertion of circuits without disassembling the signal.
- Easy migration toward higher multiplexing orders, given that they use the same work philosophy.
- Headers enable the enhancement of the network's operation, administration, and maintenance procedures (OAM).
- PDH g.702, ATM, etc., signals can be transported.
- Counts on protection integrated mechanisms.
- Defines an open optic interface to allow interconnection with other equipments.

3.2.1.10.2 Transmission Rates

The basic SDH unit is the STM-1 structure. Four STM frames are united in two fields or multiplexed to give an STM-4, which has a higher transmission rate. STM-16 and STM-64 offer higher transmission rates and withstand a greater number of signals in their area of useful load. Thus, STM-4, STM-16, and STM-64 can be seen as thicker pipes.

3.2.1.10.3 Section Overhead

Data bytes are added to the STM structure, providing a communication channel between adjacent nodes habilitating the transmission control over the link. This allows for both nodes to "speak" between each other in the event of section failure, for example, whenever there is a protection switching.

A *path* or a *route* is a term used when referring to a point-to-point circuit for traffic; in other words, this is the trajectory followed by a virtual container through the network. A *section* is defined as the transportation link between two adjacent nodes. A path is made up of a specific number of sections.

Using the initial pipe analogy, the section can be seen as the length of a pipe between two network's nodes and the path as the route taken by the virtual containers on these piping sections.

The end users' traffic will be transported in virtual containers using a given path over several sections. (This is a simple and introductory definition. Actually, paths and sections are different layers of the transportation network, as will be described later.)

An STM is dedicated to a single section; therefore, the section overhead is processed in each node and an STM with new overheads is built for the following section. However, the virtual container follows a path over different sections, so the path overhead remains with the container from one end of the path to the other.

The data goes into the network as 2 Mbps digital flows that will be adjusted in VC-12 virtual containers. An SDH element will multiplex this signal together with the other three tributary signals in an aggregated signal of higher transmission rate. In the example, this is an STM-1 signal of 155 Mbps and it is in the local SDH network. This signal can then be multiplexed so as to give an STM-4 signal at 622 Mbps in the following level, reaching STM-64 when carried at 10 Gbps. Many signals are transported in one fiber in this flow of higher transmission rate, which is known as a *main* or *backbone* network and will transport the data to a given geographic site.

The 2 Mbps signal may be extracted and delivered to its destination, or if its destination is a terminal equipment, the aggregated signal is demultiplexed descending to the 2 Mbps signal. The SDH multiplexing structure defines the standard path to map the signals included in an STM, whose basic unit is an STM-1 structure (155 Mbps). The value of other basic transmission rates is defined by using a multiplication factor of 4. These are the 622 Mbps known as STM-4, 2.5 Gbps known as STM-16, and the 10 Gbps as STM-64.

But, why do we increase the transmission rate from STM-1 to STM-16 or STM-4? Transporting data from one point to the other requires an optic fiber located between one place and the other. This is indeed a costly installation; therefore, the number of installed fibers is limited, with each fiber carrying as much data as possible, which is feasible by transporting the data at a higher transmission rate, as is the case in STM-64.

3.2.1.11 Elements of Synchronous Transmission

SDH transmission equipment has three basic functions: multiplexing, line termination, and cross-connecting. In the past, these functions were provided by pieces that were independent of and different from the equipment, but with the introduction of SDH, it is possible to combine such functions into a simple network element.

3.2.1.11.1 Functionality of a Network Element

Multiplexing is the combination of various low-speed signals into a single high-speed signal, which achieves maximum use of the physical infrastructure. Synchronous transmission systems use Time-Division Multiplexing (TDM).

In line termination/transmission, in one direction, the digital tributary signal is terminated, multiplexed, and transmitted in a higher speed signal. In the opposite direction, the higher transmission rate signal is terminated, demultiplexed, and reconstructed as the digital tributary signal. This is the task of line terminals. Synchronous transmission networks typically use fiber optics as physical transportation links, so this requires termination and transmission of optic signals.

In PDH systems, the termination, multiplexing, and transmission tasks require different, stand-alone equipment modules, but in SDH these functions may be combined into a single network element.

The use of cross-connects in a synchronous network implies establishing semi-permanent interconnections between different channels in a network element. This allows traffic to be sent at the virtual container level. If the operator needs to change the network traffic circuits, forwarding may be achieved by changing connections.

This description might suggest that a cross-connect is similar to a circuit switching, but there are basic differences between them. The main difference is that switching works as a temporary connection that takes place under control of an end user, whereas a cross-connect is a transmission technique used to establish semipermanent connections under the control of the operator through its network management system. The operator will change this semipermanent connection as the traffic pattern changes.

Other terms used in the functionalities of SDH network elements are *consolidation* and *aggregation*.

Consolidation takes place when traffic in partially used routes is reorganized in a simple path with a higher load of traffic density.

Grooming takes place when incoming traffic that is aimed at various destinations is reorganized. Traffic for specific destinations is reorganized in paths with other traffic for such destinations. For example, a specific type traffic like ATM or data traffic with different destinations may be separated from Public Switching Telephone Network (PSTN), or Switching Telephone Network, traffic and be carried on a different route.

3.2.1.12 Types of Connections

Different types of connections between elements may be established, namely the following:

- Unidirectional—a one-way connection through the SDH network elements; for example, sending traffic only.
- Bidirectional—a two-way connection through the network elements, having information delivery and receipt functions.
- Drop and continue—a connection where the signal is downloaded to a tributary of the network element, but it also continues via the aggregate signal toward another network element. These connections may be used for broadcasts and protection mechanisms.
- Broadcast—a connection where an ingress virtual container is carried to more than one egress virtual container. In essence, a signal entering the network element may be transmitted to several places from the virtual container. This type of connection may be used for video broadcasts, for example.

3.2.1.13 Types of Network Elements

The recommendation of ITU-T-G.782 identifies examples of SDH equipment through a combination of SDH functions. They are classified as multiplexers (of which there are seven variances) and cross-connects (where there are three variances). To simplify, only three types of SDH network elements will be considered: line systems, ADMs, and Digital Cross-Connectors (DXCs).

3.2.1.13.1 Line Terminals

Line terminals are the simplest SDH network element. They will implement only line termination and the multiplexing function, so their use is typical in point-to-point configurations. Certain tributary flows will be combined in the line terminal to generate a higher speed aggregate flow and this will be transmitted to an optical link. Network elements are required at the two end points of this link and a fixed client circuit connection is established between these two terminal points.

3.2.1.13.2 Add-Drop Multiplexers (ADMs)

These devices provide the cross-connection function together with line termination and multiplexing. In SDH it is possible to drop a virtual container and add another container to the STM signal in the opposite direction directly without having to ruffle it as seen previously. This basic advantage of synchronous systems means that it is possible to flexibly connect signals between network element interfaces (aggregate or tributary). This routing capability allows the cross-connection function to be distributed by the network, with better results than concentrating it in a large dedicated cross-connector.

3.2.1.13.3 Dedicated Cross-Connectors (DXCs)

As described earlier, cross-connectivity of ADMs allows for the cross-connection function to be distributed through the network, but it is also possible to have a single cross-connect equipment. Digital cross-connectors (DXCs) are the most complex and expensive SDH equipment.

It is not the inclusion of blocks with cross-connection functions that distinguishes DXCs from ADMs, but the presence of supervision in higher to lower order connections. This capability of supervising connections is DXCs' distinctive characteristic.

3.2.1.14 *Configuration of an SDH Network*

- Add-Drop Multiplexer—allows insertion and elimination of the SDH signal
- Terminating unit—similar to Add Drop, but is designed to terminate with an SDH signal
- Digital cross-connecter—may provide a link between different level STM signals
- Repeater—amplifies the SDH signal
- ITU-T SDH standards—ITU's telecommunications sector (ITU-T) is responsible for coordinating and developing SDH standards for the world. Table 3.4 lists the most important SDH standards.

Table 3.4

Standard	Description
ITU-T G.707	Network node interface for SDH
ITU-T G.781	Recommendations structure for SDH
ITU-T G.782	Equipment types and characteristics for SDH
ITU-T G.783	Characteristics of functional blocks for SDH
ITU-T G.803	Architecture of transportation networks based on SDH

About Asynchronous Transfer Mode (ATM) we can mention the following characteristics: It is a switching protocol for small packets of data units—with a fixed size of 53 bytes—called *cells*. In an ATM network, communications are established through a set of intermediate devices called *switches*. ATM is the complement of STM, which means Synchronous Transmission Mode. STM is used in telecommunications networks to transmit data and voice packets over long distances. The network is based on switching technology, where a connection is established between two points before data transmission begins. This way, the end points identify and reserve a bandwidth for the complete connection. An ATM cell consists of small, 53-byte-long (five header and 48 information) packets. The cell header consists of the routing information, cell priority, and quality of service, and verifies validity of the header. A bit, which is found in the header control field, is specified. The header control field also contains a bit that specifies whether the packet is a control packet or a regular packet, and has another bit to indicate whether the packet can be eliminated or not in case of congestion. The cell size derives from a commitment between a series of desirable characteristics for each type of traffic. On the one hand, for transmission efficiency reasons, it is convenient that cells be reasonably large. On the other hand, from a data transmission point of view, it is also advisable that cell size be large to prevent excessive segmentation.

The following are some characteristics of ATM networks:

On-Demand Bandwidth. Bandwidth assignment is performed as a function of the traffic forwarding demand.

Packet Switching Operation. By using fixed-length packets one can use switching nodes at very high speeds.

When two ATM nodes want to transfer, they must first establish a channel or connection by means of a call or signaling protocol. Once the connection is established, the ATM cells include information that allows identification of the connection to which they belong.

Scalability. ATM is designed for a wide range of speeds and different types of media.

Designed for LAN and WAN.

Originally Designed for Use Over Fiber Optic Lines. Availability of copper-based standards.

Designed for High-Reliability Networks. Technologies such as fiber today provide an error rate between 10 and 12 and between 10 and 15. For this reason, network protocols may be downloaded from error control and flow control functions. This can be done end-to-end by the higher layer protocols in the terminal equipment.

Two Defined Levels. The ATM Adaptation Level (AAL) and the ATM level. The AAL is responsible for carrying out the service convergence (data, voice, and video) process and the fragmentation and reassembly process, because application data comes in the form of packets and, as we have mentioned, ATM operates through cells. The ATM level is responsible for multiplexing cells and performing switching through the VPI/VCI values established for the different virtual circuits.

In this section we have presented more characteristics of SONET and SDH compared to ATM; this is because ATM, as well as MPLS, is increasingly falling into disuse, while SONET and SDH continue being used.

Having made some comments about SONET, SDH, and ATM technologies, we will now specify the connectivity and design scheme.

The carrier backbone design is shaped as a loop and takes place through the ADMs, over which, as its name says, one can multiplex voice and data services. Traditionally, this connection with ADMs takes place through ports OC-48 in SONET or STM-16 in SDH, with a transfer rate of 2.4 Gbps, or it can also take place through ports OC-192 or STM-64, with a transfer rate of 9.6 Gbps, in both cases without using DWDM, which will be discussed later. The ATM switches are connected to these ADMs to perform the switching function. In certain cases both devices may be integrated into a single device. Figure 3.22 shows a connectivity scheme with ATM/SDH or SONET in the carrier backbone, and integration with different last-mile technologies such as Ethernet (LAN 1), WiMAX (LAN 2), Frame Relay (LAN 3), and ADSL (LAN 4). Other types of technologies may also be connected. In this loop we can see three ADMs and three ATM switches connected to the ADMs through SONET or SDH. In this example, three multiplexers and three switches are specified, but in practice three ADMs and, for instance, only two ATM switches to perform the switching function could be installed. The foregoing depends on the network's switching requirements.

Figure 3.22 also shows that there is distribution equipment to the last mile. This is the case, for example, of connections with WiMAX, Frame Relay, and ADSL. In this case, this distribution equipment connects through SONET or SDH to the ADMs. The last mile via Ethernet connects directly to the ATM switch. What happens is that this equipment is multi-platform equipment, and just as it supports

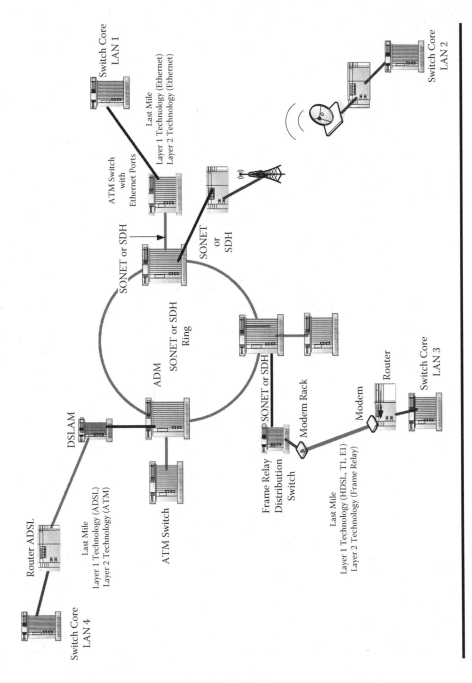

Figure 3.22   ATM backbone.

SONET or SDH technology through certain cards, it can also support Ethernet through others.

Now, in cases where ATM is used as a switching technology, virtual circuits will be configured through the VPI/VCI and, in cases where there are zones with Frame Relay, the DLCI will be used. Figure 3.23 shows a scheme in which a virtual circuit is defined that connects to LANs 3 and 4. A virtual circuit can be defined as follows: The last mile of base 3 to the carrier takes place through Frame Relay and in this case a value of DLCI 20 has been defined. Then, the circuit continues in ATM between the distribution equipment with the ADM, and in this case, the value of VPI/VCI 1/100 has been defined, which reaches the ATM switch through the loop. Next, it continues in ATM through the SDH loop to the next ATM switch in the direction of LAN 4 using VPI/VCI 2/200. Subsequently, it switches direction toward DSLAM via the SDH loop with value VPI/VCI 1/150 and, finally, ends in ATM in the last mile between the DSLAM and the router located in LAN 4 through an ADSL line with value VPI/VCI 4/200. In this case, we have been able to analyze an end-to-end of the virtual circuit connecting the last miles through the SDH loop.

3.2.2 Metro Ethernet

An emerging technology in the carrier network's core is Ethernet, commonly called Metro Ethernet in these terms. Ethernet has had a marked evolution since the 1980s, from being a very inexpensive network with low yield in the 1980s and early 1990s to being a high-speed network and having quality service management since the mid-1990s, to becoming a technology for access networks and even carrier cores from 2000. Traditional speeds with Metro Ethernet are 1 and 10 Gbps without DWDM or multiplexing technology, with DWDM in possible higher speeds. Figure 3.24 shows a connectivity scheme with Metro Ethernet in the carrier backbone and the integration with different technologies in the last mile such as Ethernet (LAN 1), WiMAX (LAN 2), Frame Relay (LAN 3), and ADSL (LAN 4). Other types of technology may be connected as well. With this last comment we can conclude that different technological access platforms may still exist as is done with SDH. In the loop we can see three Metro Ethernet switches connected among them and forming the metropolitan ring. Integration with different technologies is last mile, as in Ethernet (LAN 1), WiMAX (LAN 2), Frame Relay (LAN 3)—although this one is very little used worldwide—and ADSL (LAN 4). With these schemes we can show that the last miles, regardless of the type of access technology, can integrate to the core in Metro Ethernet.

3.2.3 DWDM

As we have done with the previous technologies, first we will discuss some basic information about Dense Wavelength Division Multiplexing (DWDM) technology and then we will explain the connectivity and design scheme.

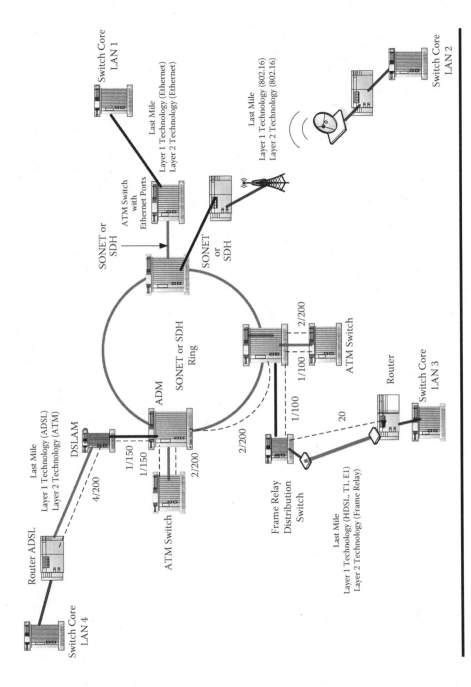

Figure 3.23 Virtual circuit.

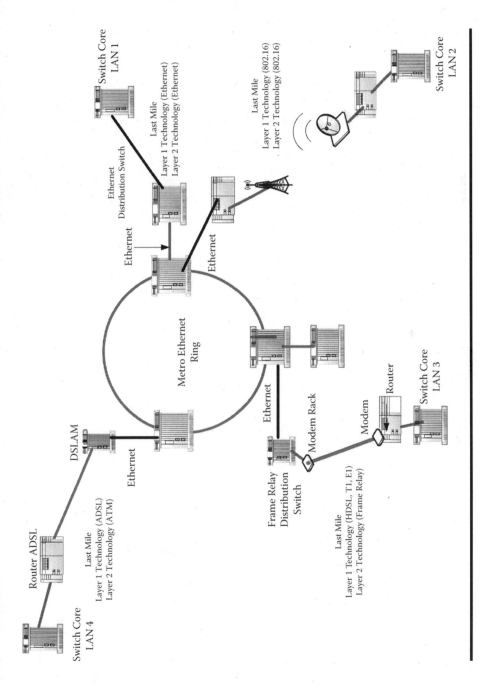

Figure 3.24 Metro Ethernet backbone.

As mentioned when explaining the DWDM technology, with DWDM technology one can increase transmission capacity over fiber optics by using different wavelengths.

DWDM is wavelength multiplexing, also known by its American denomination, Wavelength Division Multiplexing (WDM). The latter consists of sending various signals of differing wavelengths simultaneously over the same line. Wavelength multiplexing and demultiplexing are performed by means of passive optical components, similar to the decomposition and recomposition of rainbow colors through a prism. Wavelength multiplexing also opens up options of optical addressing in the networks. This way, communications may be directed for good in one or another direction according to their wavelength.

WDM provides the following advantages:

- ◼ Maximizes capacity of the fiber
  - Increases the capacity of existing optical fiber
  - Reduces the amount of new optical fiber that must be added
  - Allows gradual growth of the capacity as demanded
- ◼ Transmits a large variety of different optical signals
  - Is capable of managing different types of signals, for example, OC-48 and/or OC-192 and/or asynchronous signals, simultaneously
  - Is an independent protocol, which means that it carries only signals. It can carry FDDI, ESCON, FICON, and/or Ethernet.

WDM and Time Division Multiplexing (TDM) work jointly to optimize capacity of the fiber. TDM generates the bits flow more rapidly. This bit train, whether synchronous or asynchronous, is input to a WDM system, together with other multiplexed flows. Such flows, which originate in a TDM system, are multiplexed at wavelengths assigned for transportation over optical fiber. Every process increases the total capacity of the link.

DWDM is a technology used at the core of an optical transportation network.

The main components of DWDM can be classified by their position in the system, as follows:

- ◼ On the transmitter side, lasers with accurate and stable wavelength
- ◼ On the link, optical fiber with low losses and good transmission performance in the relevant wavelength spectrum, in addition to flat gain optical amplifiers, to amplify the signal in long distances
- ◼ On the receiver side, photodetectors and optical demultiplexers using low thickness filters or diffractive elements
- ◼ Optical Add Drop multiplexers and optical cross-connect components

These and other components, with their associated technologies, are discussed next. Although much of this information, especially the pros and cons of several competing technologies, may be more important to the system designer than to end users or the network designer. It may also interest other readers.

Having introduced DWDM, we will now specify the connectivity and design scheme.

The number of carriers who are implementing this technology in the backbone of their network is constantly increasing. DWDM can be used, for example, through Metro Ethernet or SONET or SDH; this ensures an increase in the transmission capacity of carriers in the traditional networks with SONET or SDH or with Metro Ethernet. In this example, the metropolitan ring consists of the Optical Add-Drop Multiplexer (OADM), whose functionality is to multiplex by wavelength. Figure 3.25 shows a network design with DWDM. A client's LAN 1 network is connected to the carrier through a last mile in Ethernet and subsequently this distribution equipment connects to the OADM 1 using λ_2. The LAN 2 connects through WiMAX in the last mile to the distribution equipment and this in turn connects through λ_1 to the same OADM 1. The metropolitan ring consists of the OADMs, of which OADM 1 is part, and as can be seen through the optical fiber that communicates the OADMs, multiplexer OADM 1 is transmitting through λ_1 and λ_2 ($\lambda_1 + \lambda_2$). LAN 3 network is connecting to the metropolitan ring through Frame Relay in the last mile and this distribution equipment connects to OADM 3 through λ_2. Finally, to the OADM 2 connects two distribution devices, an ATM switch, through λ_2, and a DSLAM, through λ_1. The speed of the carrier's backbone would be given by the speed of the physical port, multiplied by the number of wavelengths. For example, if the core is Metro Ethernet and the port is at 10 Gbps, one could transmit with two wavelengths, as is the case of Figure 3.25, up to 20 Gbps. In case it is SONET or SDH and a port at 9.6 Gbps whether OC-192 or STM-64 with two wavelengths, one could also transmit almost 20 Gbps. The maximum number of wavelengths available in industrial equipment until present is 192 λ, although equipment with 320 λ is soon expected.

3.3 GMPLS

The last technology associated to carriers' backbones in MAN or WAN networks we will discuss is Generalized Multiprotocol Label Switching (GMPLS), although in practice today carriers have implemented Multiprotocol Label Switching (MPLS); this means that the switching function takes place through layer 2 labels. The core in an MPLS network is made up of the Label Switched Routers (LSRs) or MPLS label switchers.

MPLS was invented in the 1990s and has taken different contributions or properties from IP switching (Ipsilon), Cell Switch Router (Toshiba), tag switching (Cisco), and IP switching based on aggregate routes, or ARIS (IBM). These companies or technologies used label switching as a method for sending data. The main idea of MPLS is to add a label to each packet to be sent. These packets are assigned a

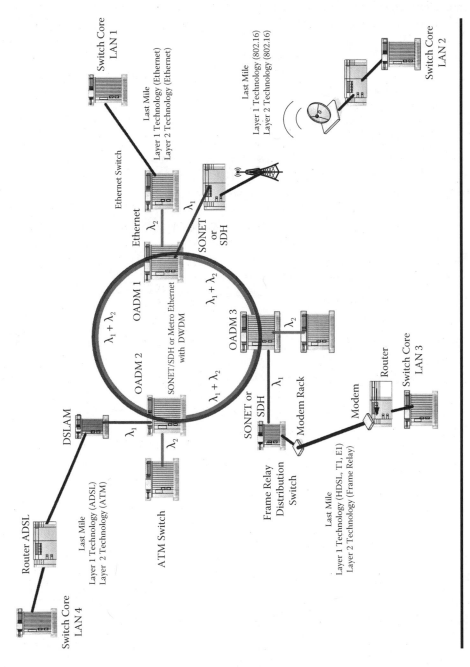

Figure 3.25 DWDM backbone.

pair of short length values that synthesize the source and destination of said packet. MPLS is a versatile solution aimed at addressing current network problems such as speed, scalability, quality of service, and application of traffic engineering. MPLS emerges as an intelligent solution for bandwidth management and service requirements, providing effective assignment, delivery, and switching of information or data through networks [DAV00]. Multiprotocol switching technologies such as MPLS separate the delivery and control characteristics or functionalities into two different, clearly identified components. The first component is responsible for the construction and maintenance of the delivery table. This delivery component is actually the one that is in charge of sending the packages. The second component is the signaling scheme to establish labels along the path between the source node and the destination node [DAV00].

3.3.1 MPLS Packet Fields

- Label—the field containing the current or real value for the label
- Exp—bits used for experimentation at the class of service (CoS) level
- S—a bit that is placed on 1 in the last label entering the stack of labels and on 0 in the rest of elements of labels
- Time to live (TTL)—field with a length of 8 bits and used to indicate the life value of the packet

3.3.1.1 Characteristics

MPLS has the following characteristics:

- Specifying a mechanism to manage traffic between varied machines, hardware, and applications
- Remaining independent of layers 2 and 3
- Providing a medium to assign or map IP addresses to labels (second-level value) used for reforwarding information or data
- Profiting of existing implementations for routing, such as OSPF (the shortest route) and RSVP (resource reservation protocol)
- Supporting in layer 2 technologies such as ATM, Frame Relay, PPP, and Ethernet, among others

3.3.1.2 Components

The MPLS for transmission or reforwarding of data is supported on a series of elements:

- Label Switched Path (LSP). LSPs are the routes through which data or the path flow from the source to the destination along which a packet is forwarded in a network. The main characteristic of LSPs is that they are formed by means of labels; that is, LSPs are but a path formed by a series of labeled nodes that allow a packet to be forwarded by label exchanges of one MPLS node to another MPLS node. An LSP in turn has two elements: the LSP Ingress (where the route begins) and the LSP Egress (where the route ends). A more formal definition may be given by the following formula: an LSP m for a specific packet p is a sequence of routers, $<R_1, ..., R_n>$, with the following properties:

- R1 is the LSP Ingress. It is an initial LSR to which is assigned the P label of the label stack. The stack has a size m.
- For every i (indicates the number of LSRs), $1 < i < n$, p has a label stack with size m when it is received by LSR R_i.
- During the exchange of the P's label of R_1 toward R_{n-1} the label stack will never have a size smaller than m.
- For every i, $1 < i < n$: R_i transmits p toward R_{i+1} by means of MPLS, using the label found on the top of the label stack.

■ Label Switched Router (LSR). LSRs are devices designed to exchange or switch labels. In essence, they continue working like traditional routers or switches with the characteristic that to carry a packet in a network from the source to the destination they use label switching.

Within a set of LSRs, there are initiation LSRs called Edge or Ingress/Egress Routers (LERs); these are label switches that do the interface between switches that do not support MPLS with those that do support this technology.

■ Label Information Base (LIB). LIBs are tables that contain the information regarding the ingress and egress labels and the port through which the information packet must be sent. Every LSR relies on a LIB to learn the path that the packet must follow.
■ Label Distribution Protocol (LDP). A Label Distribution Protocol is used in a network that supports MPLS to establish and maintain label assignment. For MPLS to operate correctly, the information regarding label distribution must be transmitted with a high degree of surety or confidence. Moreover, it is desirable that the control flow has the capability to carry multiple label messages in a simple packet.
■ Label Switched Path (LSP). Label Switched Path is the path through which the information transfer will take place. This path is created by a set of devices called LSRs that represent an MPLS domain. The path or LSP is configured for a specific packet to travel or be forwarded following the labeled trajectory.
■ Forwarding Equivalence Class (FEC). A representation of a set of packets that contain or share the same carrying characteristics or requirements. All packets that are assigned to a specific FEC are treated equally for transmission to their destination.

3.3.1.3 Operation

The following steps must be followed in an MPLS supporting network to forward a package from a source node to a destination node [ROS01; DAV00].

3.3.1.3.1 Creating and Distributing Labels

- Before starting transmission of traffic, the routers (in this case the LSRs) assign a label to a specific FEC and then the routers must construct their tables.
- In addition, with the help of LDP in a downlink way (from destination node to source node), routers begin distributing the labels and their assignment in a specific FEC.
- With the help of LDP, the MPLS traffic characteristics and capabilities are negotiated.

3.3.1.3.2 Creating the LIB in Every Router

- When a label is assigned to an LSR, the LSR creates a record in LIB to save the information regarding said label.
- The table content will specify the mapping between a label and an FEC.
- Mapping is establishing the relationship between the port and the ingress label with the port and the egress label.
- Table entries are updated regardless of the assignment occurring in the labels.

3.3.1.3.3 Creating the LSP

- The LSP is created in the opposite direction of the creation of LIB entries. In other words, the LSP is created from the destination node to the source node, whereas entries to the LIB are created from the source node to the destination node.

3.3.1.3.4 Inserting and Searching the Label in the Table

- The first router uses the LIB table to search for the next hop and the request of a label for a specific FEC.
- Similarly, the following routers use the labels to find the next hop.
- Once the packet reaches the egress LSR, the label is eliminated from LIB and is forwarded to the destination.

3.3.1.3.5 Forwarding Packets

The following example shows how packet forwarding takes place from the LER (ingress LSR) to the LER (egress LSR).

- Initially, the LER might not have an assigned label for this packet because this is the first request for delivery or the first packet to be forwarded. As in an IP network, the route will first be constructed using the IP addresses, and the LER knows that the next step or hop is the LRS.
- The request will be spread throughout the network.
- Every intermediate router (LRS_i) will receive a label in its link with a down-link router (LRS_{i+1}).
- The LER will insert the label in its LIB and will forward the packet to the next LSR.
- Every LSR in the route will examine the label that accompanies the incoming packet and will replace such ingress label in the packet with an egress label.
- When the packet reaches the final LRS, the LSR will remove the label because the packet is leaving the MPLS domain toward the destination node.

Figure 3.26 shows a connectivity scheme with MPLS in the carrier backbone and the integration with different technologies in last mile such as Ethernet (LAN 1), WiMAX (LAN 2), Frame Relay (LAN 3), and ADSL (LAN 4), just like in the previous cases. The LSPs are defined through this equipment, which defines the logical paths for data transmission and over which the quality of service parameters are established.

3.4 MAN/WAN Solution with IP

Finally, in this section we will show a design involving the previous technologies with IP. In other words, we will show a homogenous scheme in layer 3 and the routing process between LAN networks through the carrier networks. Once the IP and layer 2 are integrated, convergence of services for transmission of data, voice, and video can be performed.

Figure 3.27 shows a connectivity scheme through IP. In this case we have decided to use the GMPLS technology as layer 2. The case would be similar if we use any other layer 2 technology shown in previous sections. It could also be possible that the carrier backbone is in DWDM technology, which would imply a change in layer 1, for which the scheme currently shown would not change.

For this example, let's assume that for LAN networks 2, 3, and 4 an LSP has been created to connect these networks to LAN 1. The last mile technology notwithstanding, LSPs can be created in the carrier backbone. This is the scheme that is currently being implemented, although it is clear that one can also reach the last mile with the LSP. In this first case we will use private IP addresses and, later, we will show the scheme with public IP addresses, that is, we would be connected to the Internet directly.

For this example we have used IP address 10.0.0.0 as the IP address of the different LAN terminals and the WAN connections. The address 10.1.0.0/16 has been

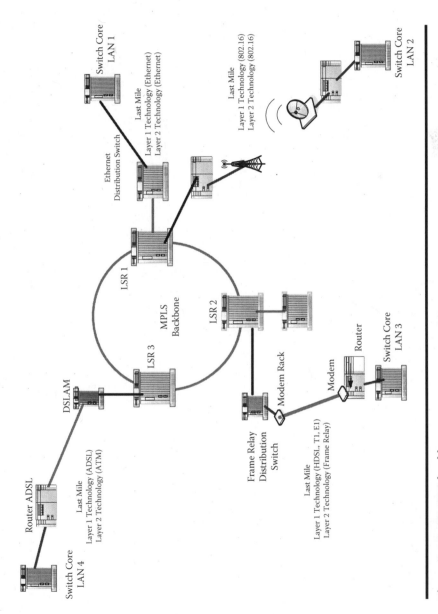

Figure 3.26 GMPLS backbone.

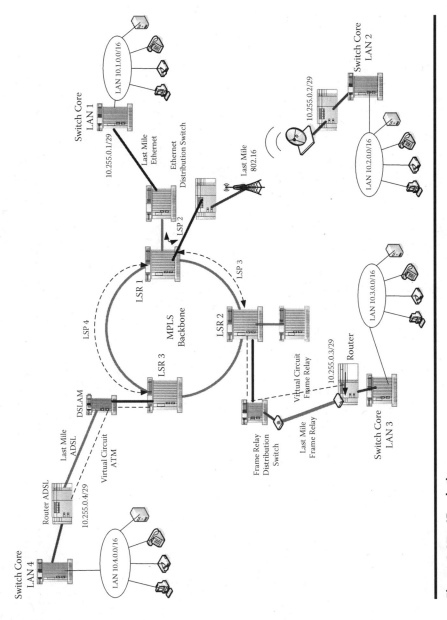

Figure 3.27 IP solution.

used by LAN 1; this means that any equipment, whether PCs, servers, routers, IP telephones, or videoconference stations, will use an address in this range. For LAN 2 the address 10.2.0.0/16 has been used; for LAN 3 10.3.0.0.0/16, and, finally, for LAN 4 the address 10.4.0.0/16 has been used. The IP addressing of the WAN network may be designed in several ways. The first way, which is shown in Figure 3.27, consists of taking one mask /29; this way we can assign six IP addresses as shown in the figure. In this case, we have selected IP address 10.255.0.0/29 for the assignment of WAN connections from the different LAN networks. This way, for example, the WAN outlet of LAN 1 has been assigned address 10.255.0.1/29, LAN 2 network has been assigned 10.255.0.2/29, LAN 3 has been assigned 10.255.0.3/29, and LAN 4 has been assigned 10.255.0.4/29. Another way to perform WAN connectivity would be assigning addresses /30 to each of the edges of the LSPs or virtual circuits. For example, the connection between LAN 1 and LAN 2 can be assigned addresses 10.255.0.1/30 to LAN 1 network and 10.255.0.2/30 to LAN 2. In the connection from LAN 1 to LAN 3 one can use the following range /30; in other words, LAN 1 would be assigned address 10.255.0.5/30 and LAN 3 would be assigned address 10.255.0.6/30.

Chapter 4

Quality of Service

In this chapter we review some basic concepts about quality of service (QoS) and the different services and configurations through the different technologies already discussed in the previous chapters. Having explained LAN and WAN network design in the previous chapters, this chapter explains the QoS configuration scheme for IP convergent applications to function appropriately. Applications and applications requirements will be discussed in Chapter 5.

The fast growing pace of network technology has raised concerns and confusion about the QoS concept and how it integrates and benefits networks [WAN01]. Communications networks users are executing critical applications within network infrastructures that allow them to perform their work efficiently and competitively. Long response times and out-of-service networks are unacceptable and may have a negative impact on the competitive business world. The cause for these variations in network performance is the best-effort type of service provided by current IP networks.

Despite the problems that arise with the best-effort model, it has been widely recognized. The real need of improving service architectures appeared in the early 1990s after results of an experimental videoconference on the Internet were analyzed. Real-time applications such as videoconferences are highly sensitive to delays and because of this do not work well on the Internet, as delays are practically unpredictable. To satisfy the requirements of these applications a new service that can provide a certain level of resource assurance to the applications is needed.

In its simplest form, quality of service is defined as the mechanism that satisfies the requirements of network applications, that is, the capability of certain network element(s) to guarantee satisfaction of network traffic needs to a certain level. All network layers come into play here. QoS assigns priority according to network traffic needs and thus manages the network bandwidth [WAN01].

The following parameters have been widely used to describe quality of service requirements [WAN01]:

■ Minimum Bandwidth. The minimum bandwidth requested by an application flow. The time interval to measure bandwidth must be specified because different intervals may yield different results. Bandwidth assignment is guaranteed by the packet-scheduling algorithms.
■ Delay. A delay request may be specified as average delay or the worst-case delay. Delay experienced by packets has three components: propagation delay, transmission delay, and queuing delay. Propagation delay is caused by the speed of light and is a function of distance, traffic delay is the time it takes to send a package in a link, and queuing delay is the waiting time that will be experienced by packets.
■ Delay Jitter. A delay-jitter request is the maximum difference between the longest and shortest delay experienced by packets. In any case, this should not be longer than the transmission in the worst case and the queuing delay.
■ Loss Rate. The rate of lost packets and the total transmitted packets. Loss of packets in the Internet is often caused by congestion, and said losses can be prevented by assigning sufficient bandwidth and traffic flow.

One of the causes of low performance in networks is congestion. Congestion is the inability to transmit a volume of information with the capacities established for a specific equipment of network. Congestion can occur because of the following:

■ Port Congestion. Multiple inlet flows compete for the bandwidth of the same outlet port.
■ Congestion in Intermediate Nodes. This happens if the backplane (switching matrix) bandwidth of a node is lower than the total aggregate of its inlets.
■ Network Congestion. This happens if at any point between the source and destination, one or several network equipments or links experience congestion.

There are two options to solve this problem. The first is to increase network bandwidth, which has a cost and the bandwidth is not infinite. The second is to manage the available bandwidth intelligently. The network can monitor use of its bandwidth, observe the symptoms that signal congestion, and reinforce policies relating to the provision, use, and distribution of the available bandwidth.

Why is QoS necessary? Congestion is the main cause of failures and of long response times in networks when one accesses applications sensible to congestion. QoS solves this problem by providing network resources so that the most important applications acquire bandwidth at the expense of those with lesser importance.

To solve congestion problems, a network administration scheme that is capable of detecting and reacting to congestion in order to control it intelligently is necessary. Such a scheme is called *traffic control*.

Certain applications do not allow delay fluctuations, among them UUCP, TN3270, video transfer, and VoIP. This delay can be introduced by network equipment, aggregate link, a protocol, or a server.

We will talk about quality of service when the network infrastructure has to deliver a service level to a specific application, meeting its performance requirements. QoS can distinguish between different types of traffic and assign resources according to parameters such as bandwidth, jitter (delay variation), and package losses [WAN01].

Grouping similar types of traffic within one class allows one to use wider QoS policy schemes. These classes are known as *class of service* (CoS).

The term *soft QoS* is used when QoS is applied without end-to-end signaling and the bandwidth is managed by independently established policies for each interface or in each intermediate node in the network. This option manages congestion based on priority for each class of service, but does not provide absolute end-to-end guarantee. To establish a global significance, QoS rules and mapping are distributed by means of network administration platforms. Some examples of this model include differentiated services (DoS), IP precedence, ToS, and priorities based on tag 802.1P/Q.

Hard QoS is used when the parameters (bandwidth, delay, etc.) can be negotiated end-to-end. CoS is also used here as a way of grouping sessions with similar characteristics, but use of the session is monitored at every hop and is forced to comply with the parameters of QoS that have been negotiated. Indeed, the session creates an end-to-end pathway through the network with a single distinction (flowspec) based on a single session attribute (filterspec). This combination of flowspec and filterspec is called *flow* and is the basis of RSVP and of integrated services (IntServ).

One of the aspects in networks to comply with Service Label Agreement (SLA) is control of congestion. To perform congestion control, one of the factors to take into account is queuing management. In this case, the number of available outlet port queues has a strong impact on the capability of networks to deliver the desired QoS for network applications, as it determines how many classes of traffic can be managed. Each application group is assigned a CoS profile that directly reflects in an outgoing queue at every port of the node facilitating provision of the end-to-end service level committed in the SLA. If the number of queues is low, they may not be enough to manage the varied network applications.

The size of queues is the number of packets that it can store. Generally one thinks that "the more, the better"; however, voice and video have different requirements. In package-based systems, jitter is directly proportional to the size of the queue. Buffers that are too long may dramatically increase network jitter (for example, 10 ms for voice and 100 ms for video are acceptable values) when congestions peaks appear.

When an outlet port is congested, packets must be discarded or stored temporarily until the outlet port can send it. Once packets have fallen into one or more queues an algorithm is used to establish how the packets are transmitted from the queues to the outlet port [WAN01].

The following are some of these algorithms:

First In First Out (FIFO). This is the simplest queuing and transmission model. Packets received are placed in a single queue in receiving order and then are transmitted in the link according to the order in which they were received. This is a very simple technique, but it does not have the capability to prioritize packets or to differentiate service classes.

Priority Queuing. This is a more advanced and versatile algorithm that uses the concept of a task distributor, which is capable of intelligently decongesting multiples queues. Every outlet queue is assigned a priority level that determines the outgoing sequence. All packets are sent from high to low priority.

Programmable Bandwidth Queuing (PBQ). This is an algorithm that establishes the bandwidth that will be available for every queue from a specific port in a flexible and easy to use way, by having 1 Mbps granularity over a 1 Gbps link (0.1% granularity). This concept is similar to Weighted Fair Queuing (assignment to queues is a fixed percentage of the port's bandwidth) except that fixed binary weights and imprecise percentage restrictions are eliminated. When congestion exceeds the queuing storage factor, packets trying to be transmitted are discarded. There is an improvement of discarding, known as *selective discarding*, which is capable of differentiating traffic into two categories, so that under congestion, network equipment can decide to send some segments while discarding others. This traffic differentiation is based on the classification of segments as discard eligible or not. Having a QoS system that responds to congestion control is only an initial step because these techniques stop being effective when congestion continues for long periods of time. It is important to use congestion prevention techniques that feed the generating sources with information regarding network congestion for them to reduce the speed of traffic sent and eventually make congestion in the network disappear. These techniques include, among others, IEEE 802.3x, RED, WRED, and ECN.

Random Early Discard (RED). Its primary use is to provide flow control through the TCP congestion avoidance window, eliminating low-performance situations caused by the TCP global synchronization effect. This effect is produced in congestion situations where all new packets in the queue are discarded and the majority of sessions reduce their flow through the TCP congestion avoidance algorithm. Thus, outlet queues are emptied, even if almost empty, and congestion is eliminated. However, when flow of sessions starts to increase (via the TCP window algorithm), congestion occurs once again as well as large-scale discard of packets. In this case, the network remains between 0% and oversaturated all the time, which results in low performance of the network. This anomalous situation is eliminated by the network by making level marks on the queues. Every queue has a maximum (max) and a minimum (min) within which the RED algorithm is invoked. When traffic is below the minimum level, no package is discarded, but when packets are above the maximum level, all packets are discarded. When packets are between maximum and minimum level, RED starts to function and randomly discards packets at a variable speed (statistically, more packets will be discarded from sessions that consume more bandwidth). The discarding speed is given by the level of the queue between the maximum and minimum marks. The higher the queue level, the faster the discarding speed, and more TCP sessions will reduce their transmission speed. The resulting effect is that congestion never lasts a long time to the extent that TCP flow control mechanisms manage and prevent it.

Weighted RED (WRED). WRED uses the RED algorithm, but instead of discarding any traffic without looking at its content, discarding takes place after a classification result.

Explicit Congestion Notification (ECN). ECN is an emerging standard in IETF to mark packets with a congestion status to notify of the congestion forward. Stations that detect the congestion status will reduce the flow of their transportation sessions (e.g., TCP window), helping to control congestion.

Bandwidth Control. This essentially consists of limiting and shaping the bandwidth. This is a necessary characteristic if a QoS provided bandwidth to its SLA. Certain switches provide bandwidth control in two ways: a policy based on flow that is provided over the inlet port and a shaping policy over the outlet port. Bandwidth policies are mechanisms that limit inlet traffic over a specific port. Such policies are based on a 0.1% granularity. In excess cases, the discarding process is performed by means of the leaky model, which uses a time measure to indicate when queue traffic can be transmitted to control speed.

Traffic Shaping. The capability to liberate bandwidth at an outlet port with a regular and predictive speed. Traffic shaping provides a QoS system with important characteristics: it removes excess traffic and delivers traffic at constant speed to the next nodes.

4.1 LAN Solution

In this section we analyze quality of service associated to certain LAN technologies. The most used presently, Ethernet, will be analyzed in this case.

Traditionally, quality of service in Ethernet can be configured in two ways. The first way is by assigning priority to VLANs, and the second way is by assigning priority through a field in the Ethernet frame header called IEEE 802.1p standard. There are other ways of configuring quality of service in LAN networks, such as through TCP or UDP ports, or at an applications level, but these last schemes will be discussed later in this chapter.

4.1.1 VLAN Priority

As mentioned in previous chapters, one way of dividing a LAN network is through VLANs. This is primarily done to distribute the broadcast domains, but can also be defined to assign different priorities among such VLANs. For example, to transmit critical services such as voice or video, it is necessary that the broadcast receiving such services does not exceed 5%. It is for this reason that such services must be assigned within a special VLAN. If these critical services have already been located in a special VLAN, the following step would be to assign them a specific and higher priority than any other data exclusive service, and that is not network control.

Figure 4.1 shows a logical scheme of a VLAN configuration with different priorities. As can be observed, the highest priority in this configuration (with a value of 6) is associated to VoIP and videoconference over IP services. The following priority, with a value of 4, is associated to the corporate server's VLAN such as the application server, database, and file servers. With priority 3 is the equipment belonging to the VLAN connecting to the WAN network such as e-mail server, Web server, and the connection router. Finally, associated to priority 0 (the lowest priority) is the VLAN associated to the organization's PCs. Equipment in LAN networks traditionally manage level 8 priority, which are directly associated to the 3 bits of standard IEEE 802.1p. The lowest priority level is 0 and the highest is 7.

Figure 4.2 shows the logical scheme of the logical design illustrated in Figure 4.1. In this figure we see that the equipment belonging to different VLANs are connected to the core equipment.

The way of configuring VLANs' priority very much depends on the specific brand of the equipment. Following, we show a configuration example performed with Alcatel's OmniStack and OmniSwitch equipment for the case illustrated in the previous figures. These configurations would be done at the LAN network core switch.

As a first case, we show the example of the VLAN 1 configuration, that is, the corporate server's VLAN.

```
GROUP Number ( 1)                        :     1
Description (no quotes)                   :     Server Vlan
Enable IP (y)                            :     y
IP Address                               :     10.0.0.254
IP Subnet Mask (0xff000000)              :     255.192.0.0
Disable routing? (n)                     :     n
Enable NHRP? (n):
IP RIP mode {Deaf(d),
Silent(s),
Active(a),
Inactive(i)} (s)                         :     a
Default framing type {Ethernet II(e),
fddi(f),
token ring(t),
Ethernet 802.3 SNAP(8),
```

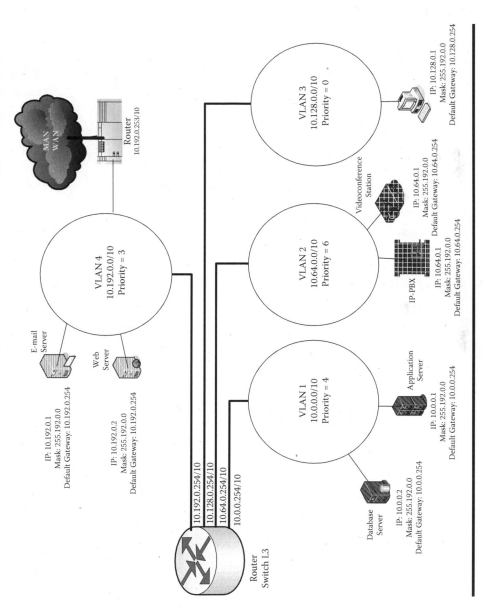

Figure 4.1 Logical scheme of VLANs with QoS.

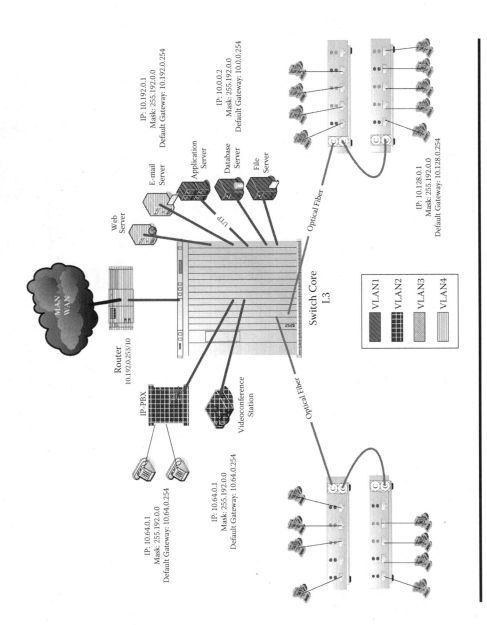

Figure 4.2 Physical scheme of VLANs with QoS.

```
source route token ring(s)} (e)              :      e
Enter a priority level (0...7)(0)             :      4
```

As a second case we show the example of the VLAN 2 configuration, that is, the VLAN of VoIP services and videoconference over IP.

```
GROUP Number ( 2)                    :      2
Description (no quotes)               :      Voice and Video Vlan
Enable IP (y)                        :      y
IP Address                           :      10.64.0.254
IP Subnet Mask (0xff000000)          :      255.192.0.0
Disable routing? (n)                 :      n
Enable NHRP? (n) :
IP RIP mode {Deaf(d),
Silent(s),
Active(a),
Inactive(i)} (s)                     :      a
Default framing type {Ethernet II(e),
fddi(f),
token ring(t),
Ethernet 802.3 SNAP(8),
source route token ring(s)} (e)  :      e
Enter a priority level (0...7)(0):   6
```

As a third case, we show the example of the VLAN 3 configuration, that is, the VLAN of the LAN network PCs.

```
GROUP Number ( 3)                    :      3
Description (no quotes)               :      PC Vlan
Enable IP (y)                        :      y
IP Address                           :      10.128.0.254
IP Subnet Mask (0xff000000)          :      255.192.0.0
Disable routing? (n)                 :      n
Enable NHRP? (n) :
IP RIP mode {Deaf(d),
Silent(s),
Active(a),
Inactive(i)} (s)                     :      a
Default framing type {Ethernet II(e),
fddi(f),
token ring(t),
Ethernet 802.3 SNAP(8),
source route token ring(s)} (e)  :      e
Enter a priority level (0...7)(0):   0
```

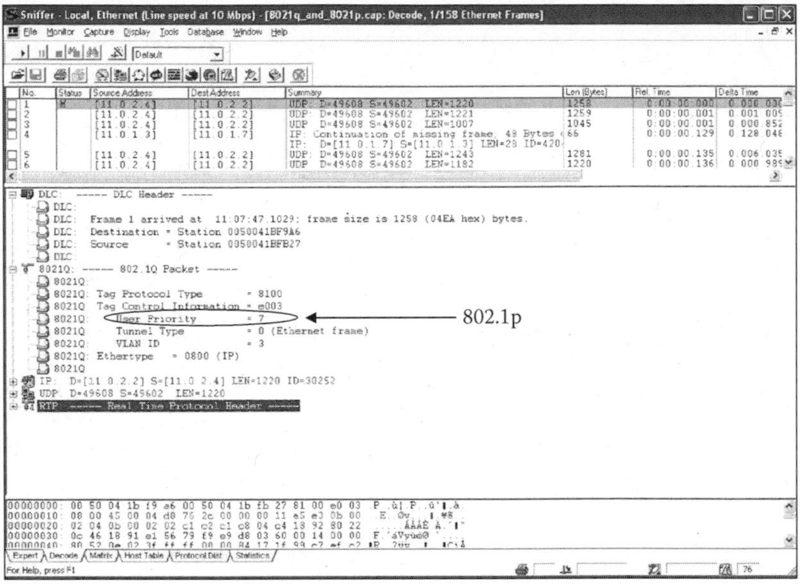

Figure 4.3   Frame tagged with VLAN priority.

Last, we show the example of the VLAN 4 configuration, that is, the external connection.

```
GROUP Number ( 4)                              :     4
Description (no quotes)                         :     External Vlan
Enable IP (y)                                   :     y
IP Address                                      :     10.192.0.254
IP Subnet Mask (0xff000000)                     :     255.192.0.0
Disable routing? (n)                            :     n
Enable NHRP? (n) :
IP RIP mode {Deaf(d),
Silent(s),
Active(a),
Inactive(i)} (s)                                :     a
Default framing type {Ethernet II(e),
fddi(f),
token ring(t),
Ethernet 802.3 SNAP(8),
source route token ring(s)} (e)                 :     e
Enter a priority level (0...7)(0                :     3
```

When VLANs are defined with these priorities, switches record the priority through the IEEE 802.1p in the Ethernet header and also record the value of the VLAN (ID VLAN) in the frame. Figure 4.3 shows an Ethernet frame with both fields.

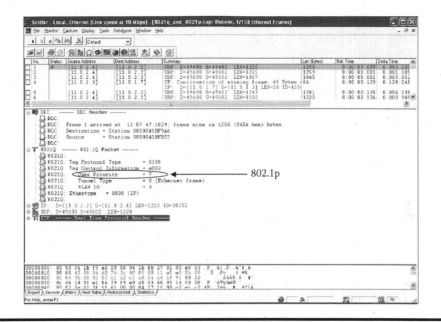

Figure 4.4   IEEE 802.1p.

4.1.2 IEEE 802.1p

Although through priority configuration in VLANs it is possible to mark the priority in the Ethernet header through standard IEEE 802.1p, this is not the only way to perform it. For example, it is possible to define the priority level at each physical port of the switch by identifying a data flow, whether TCP or UDP, or through an IP address, etc. In conclusion, we can say that field 802.1p is the marking over Ethernet to subsequently give these packets special treatment, depending on their priority value. Figure 4.4 illustrates the package with the final marking with the assigned 802.1p value. The purpose of this book is not to specify quality of service configurations in different brands, but merely to show some examples. For further detail on the different brand configuration schemes, you can consult the equipment manufacturers' reference manuals. As a sample configuration Table 4.1 shows an example of QoS in a port of a Cisco device, and from this configuration one can mark the frames in the 802.1p field.

4.2 MAN/WAN Solution

In this section we discuss management of QoS in MAN and WAN networks both for last-mile access and in the backbone technologies in the carrier.

Table 4.1

Step	Detail	Command	Example
1	Enter into the global configuration mode	Configure terminal	Configure terminal
2	Enter into the interface to configure	Interface *interface*	Interface fa0/1
3	Establish priority to the port		

4.2.1 QoS in Frame Relay

This technology provides parameters that are used to manage traffic congestion. It is possible to eliminate bottlenecks in Frame Relay networks with high-speed connections to the main network and low connections to the edge networks. Bandwidth values can be configured to limit the rate at which data will be sent from the virtual circuit (VC) in the main network. Table 4.2 shows the main concepts to be applied in Frame Relay circuits.

When a virtual circuit for voice is configured, the data can suffer as a result of the good quality of voice. The following tips should be taken into account when configuring circuits for voice:

- Do not exceed the PVC's Committed Information Rate (CIR). Because voice cannot tolerate much delay, queuing of voice packets must be minimized. When the CIR is exceeded (PVC CIR, not the CIR configured in the router), depending on how congested the rest of the Frame Relay network is, packets start queuing and, consequently, delaying. With time, queues of the Frame Relay switch have withstood enough to produce BECNs; as a result, voice quality is reduced. By maintaining the values at or below the PVC's CIR value, transportation of voice works consistently.
- Do not use Frame Relay adaptive shaping. Frame Relay adaptive shaping is used when one wants to manage bandwidth according to traffic movement, that is, whether BECNs are generated or if network congestion is detected. If the CIR configured within the map class is the same as the true PVC CIR, there is no need to destroy the traffic produced by the BECNs. If CIR is not exceeded, no BECNs are generated.
- If the Bc is small, the Tc is small (Tc = Bc/CIR). The minimum value accepted for the ideal voice quality is 150 ms. With a small Tc value there is no risk for long packets. Large Tc values can result in large spaces between the packets sent, because the router expects a complete Tc period to construct additional

Table 4.2

Term	Definition
Committed Information Rate (CIR)	The rate (bits per second) that Frame Relay providers guarantee for data transfer. The value of CIR is given by the Frame Relay service provider and configured by users in the router. Note: This rate is given during a committed time.
Amount/Number of Committed Information (Bc)	Maximum number of bits that the Frame Relay commits to transfer over a committed time interval. Tc = Bc/CIR
Committed Time (Tc)	The time interval during which users can only send Bc + Be bits. Tc is calculated as Tc = Bc/CIR. The value of Tc is not directly configured in Cisco routers; it is calculated after the Bc and CIR values are configured. Tc cannot exceed 125 ms.
Explicit Congestion Notice from the Source Router (BECN)	A bit in a Frame Relay package header that indicates network congestion. When a Frame Relay switch recognizes congestion, it updates the BECN bit in the packets with destination of the source router, telling it to reduce the transmission rate.

values to send the new package. Configure Bc = 1000 bits in the router using Tc = 150 ms.

■ Make Be = zero. To ensure that the CIR value is not exceeded, Be is set at zero; therefore, there is no excess burst.

Figure 4.5 shows a configuration of a virtual circuit in Frame Relay that enters through physical port 5 with a DLCI value of 20 and egress via physical port 6 with a DLCE value of 21. The virtual circuit is defined in both directions; that is, from physical port 6 to 5, the same DLCI values are retained. The same figure shows that the virtual circuit is defined as Best Effort in the Service Category field. This means that this virtual circuit is defined to transmit data. If the virtual circuit were to transmit voice, it should be configured as Real Time in option F6-SERV_ CAT, thereby reducing delay in the transmission of packets in this circuit because it would manage a priority higher than the frames being transmitted by a Best Effort virtual circuit. Figure 4.5 shows a configuration that can be done with a Frame Relay Mainstreet 3600 switch.

Figure 4.6 shows a virtual circuit configured as Real Time.

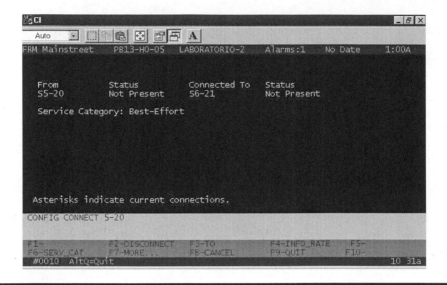

Figure 4.5   Virtual circuit in Frame Relay best effort.

Figure 4.7 illustrates a possible configuration of a virtual circuit. In this case CIR = 64 Kbps, Bc = 64 Kb, and Be = 0 Kb. The foregoing suggests that the virtual circuit ensures transmission of 64,000 bits in 1 second, which is the reason that the complete transmission is ensured and is a typical voice transmission case. Were the

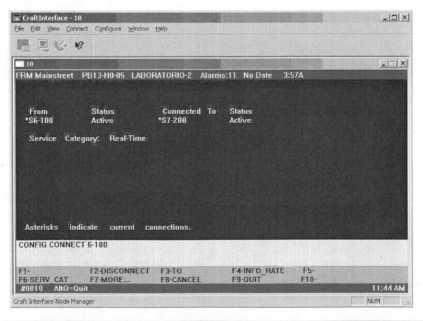

Figure 4.6   Virtual circuit in Frame Relay Real Time.

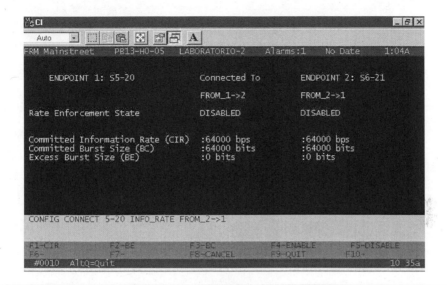

Figure 4.7 Parameter configuration in Frame Relay.

virtual circuit to transmit data, one could configure a Be value other than 0, so as to transmit information by excess. Although it has been said that not every virtual circuit for data transmission should have a Be value other than 0, for a voice transmission circuit the value of Be has to be 0.

Now, if we have VoIP at the ends to transmit over this virtual circuit, we would then have a configuration similar to the following example. For this example we have used a Cisco 2800 router. In the example we are defining in a serial (WAN) interface, a transmission through a virtual circuit with DLCI 20. In this interface the map_voice class is defined to define the parameters of the virtual circuit in the customer border equipment in the last mile. Subsequently, over the router, one defines telephone extension 100 in port FXS 3/0 through POTS service number 10. Next, in the router, one defines external extension 200 through VoIP service number 11 and indicates that in case of wanting to establish a phone call to number 200, one will have to route to IP address 200.20.33.5 through the *session target* command.

The following is the configuration of the router connected in the virtual circuit with DLCI 20:

```
interface Serial0/0
 ip address 200.20.30.1 255.255.255.0
 encapsulation frame-relay
 no fair-queue
 frame-relay interface-dlci 20
class map_voice
!
```

```
dial-peer voice 10 pots
 destination-pattern 100
 port 3/0
!
dial-peer voice 11 voip
 destination-pattern 200
 session target ipv4:200.20.33.5
!
map-class frame-relay map_voice
  frame-relay cir 64000
  frame-relay bc 64000
  frame-relay be 0
  frame-relay mincir 64000
  no frame-relay adaptive-shaping
  frame-relay fair-queue
  frame-relay voice bandwidth 64000
!
```

We have shown here how to configure virtual circuits by differentiating whether they are for data or VoIP transmission.

There is another way of associating priorities to the different services to then transmit them over different virtual circuits. A configuration scheme is shown below. In this example, four priority levels are defined. In the highest level, with a DLCI value of 20, are the Internet Control Message Protocol (ICMP) services; in the intermediate level, with a DLCI value of 21, is the Telnet service; with regular priority, with a DLCI value of 22, are the FTP service and the rest of the applications and services with low priority through DLCI 23.

```
interface Serial0/0
no ip address
encapsulation frame-relay
priority-group 1
!
interface Serial0/0.1 point-to-point
ip address 10.0.0.1 255.255.255.0
frame-relay priority-dlci-group 1 20 21 22 23
frame-relay interface-dlci 20
!
access-list 10 permit icmp any any
priority-list 1 protocol ip high list 10
priority-list 1 protocol ip medium tcp telnet
priority-list 1 protocol ip normal tcp ftp
priority-list 1 protocol ip low
```

4.2.2 QoS in ATM

As in Frame Relay networks, ATM networks manage certain parameters to establish virtual circuits with QoS. ATM defines different classes of services associated to the ATM (AAL) level. The service classes are shown in Table 4.3.

As seen in Table 4.3, one can configure virtual circuits through different service classes (CBR, VBR-RT, VBR-NRT, ABR, and UBR). Until present, the category Unspecified Bit Rate (UBR) had not been specified and, in this case, no traffic value is secured and the adaptation level that has been specified is AAL5; this is the typical Best Effort transmission in Internet.

Table 4.4 shows the parameters secured by each type of service category to comply with the service classes mentioned previously (Class A, B, C, and D).

ATM networks specify certain parameters that subsequently are applied to these virtual circuits' services categories. The following are the parameters for bandwidth:

Peak Cell Rate (PCR)—specifies the maximum rate of cells that can be introduced in the network over a virtual connection

Sustainable Cell Rate (SCR)—specifies the average rate over the number of cells that can be introduced in the network over a virtual circuit

Maximum Burst Size (MBS)—specifies the amount of time that the virtual connection can accept PCR traffic

Minimum Cell Rate (MCR)—specifies the minimum cell rate that the network must guarantee for a virtual connection

Table 4.3

	Class A	*Class B*	*Class C*	*Class D*
Time Sync	Required	Required	Not required	Not required
Bit Rate	Constant	Variable	Variable	Variable
Connection	Connection oriented	Connection oriented	Connection oriented	Connection-less
AAL	AAL1	AAL2	AAL3	AAL4
Example	Voice/video circuit emulation	Voice and video compressed	Frame Relay	Data without connections (SMDS)
Service Category	CBR (Constant Bit Rate)	VBR-RT (Variable Bit Rate–Real Time)	VBR-NRT (Variable Bit Rate–Not Real Time)	ABR (Available Bit Rate)

Table 4.4

Category	Bandwidth	Max CTD (Cell Transfer Delay)	CDV (Cell Delay Variation)	CLR (Cell Lost Ratio)	Through-put	Priority
CBR	Yes	Yes	Yes	Yes	Yes	Highest
VBR-RT	Yes	Yes	Yes	Yes	Yes	High
VBR-NRT	Yes	No	No	Yes	Yes	Medium
ABR	Yes	No	No	No	Yes	Low
UBR	No	No	No	No	No	Lowest

At the QoS level, ATM specifies the following parameters:

Peak-to-Peak CDV—specifies variation in the delay of the cells to be transmitted
Maximum Cell Transfer Delay (Max CTD)—specifies the delay of the cell from one end to the next
Cell Lost Ratio (CLR)—ratio of cells transmitted to lost cells

Next we will explain how bandwidth parameters are defined in virtual circuits through the service categories. Figure 4.8 shows that traffic is constant on CBR virtual circuits by means of parameter PCR.

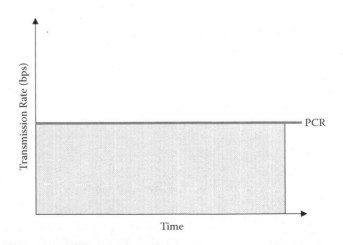

Figure 4.8   CBR virtual circuit.

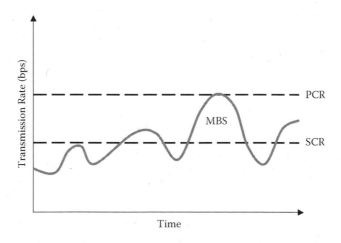

Figure 4.9   VBR virtual circuit.

In the case of virtual circuits configured as VBR (Figure 4.9), traffic is maintained by means of SCR and the maximum values are controlled through parameters PCR and MBS.

In the case of ABR virtual circuits (Figure 4.10), a minimum value is always secured by means of parameter MCR and any additional value depends on the availability of resources in the ATM network.

Finally, no traffic value is secured in a UBR virtual circuit (Figure 4.11)—the only thing that is done is control that the user does not attempt to transmit more than the PCR through the circuit—but in this type of circuit none of the values is secured. In any of the configured virtual circuits that we have discussed, we notice that the maximum transmission value is always controlled through the PCR parameter. In

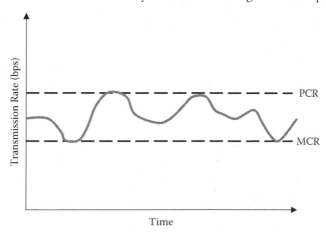

Figure 4.10   ABR virtual circuit.

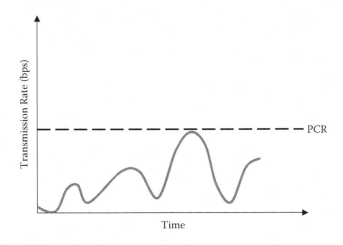

Figure 4.11   UBR virtual circuit.

conclusion, if one wants to transmit services or applications, whether with voice or video, it would be necessary to configure a CBR or VBR-RT virtual circuit.

Having seen how parameters are defined in ATM, we will show some configurations in Alcatel ATM equipment, creating some VBR-RT virtual circuits. Figure 4.12 shows the creation of a virtual circuit that enters via the SONET or SDH

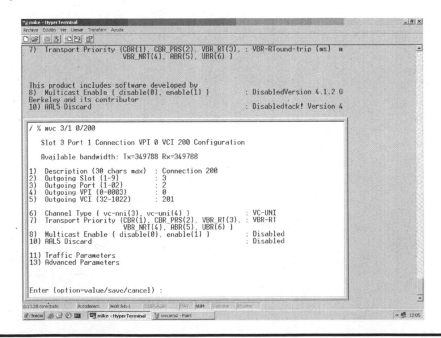

Figure 4.12   VBR-RT configuration.

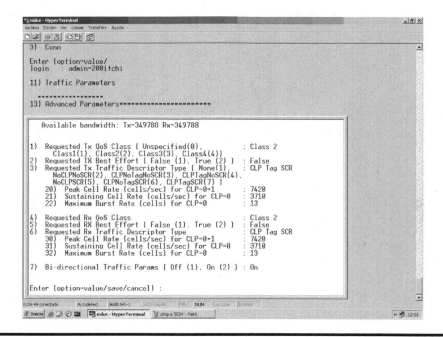

Figure 4.13   Traffic parameters.

3/1 port and egresses through port 3/2. In this example the circuit is a Permanent Virtual Circuit (PVC) with values VPI/VCI 0/200 for port 3/1 and 0/201 for port 3/2. The figure shows that the virtual circuit is configured as VBR-RT.

Figure 4.13 shows the configuration of traffic parameters PCR, SCR, and MBS. In this example PCR and SCR are in cells per second and MBS in cells. The value of PCR equal to 740 cells/s corresponds to 3 Mbps and the value of SCR equal to 3710 corresponds to 1.5 Mbps.

Figure 4.14 shows two QoS parameters: first, the value of Priority and second, the value of Cell Delay Variation (CDV). In conclusion, this shows an example of how to configure a virtual circuit with QoS to support the transmission characteristics sensitive to the consumed bandwidth and delay.

4.2.3   QoS in ADSL

Although the contribution of ADSL is associated in layer 1 of the OSI model, it is necessary to configure certain parameters in every ADSL line depending on whether one is transmitting data or critical services such as voice or video. This case demonstrates that when QoS is applied, it must be done in all the levels of a reference model and must also be end-to-end, that is, from the service transmitter through the final receiver.

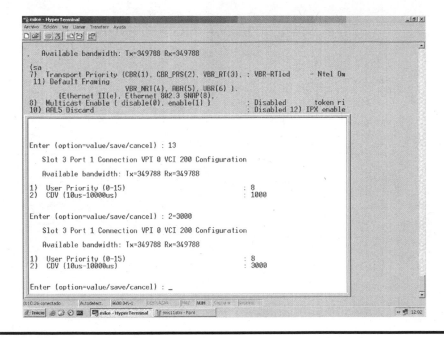

Figure 4.14 QoS parameters.

Figure 4.15 shows the different parameters that are associated to line ADSL 1/2. In this example, one is showing configuration of a DSLAM Paradyne BitStorm 4800. This figure shows the parameters associated to the uplink and downlink transmission rate. In the case of Latency, it shows that it can be configured as Interleaved or Fast. In case the line is used exclusively for data transmission, it must be configured as Interleaved. But in the case of voice or video transmission, even, for example, IPTV, this line must be configured as Fast.

In Figure 4.16 we discuss the Behavior parameter. This parameter can be configured in the DSLAM as Adaptive in case services with variable transmission rates, such as circuits VBR or ABR in ATM and even UBR, are needed. If services with constant transmission rates, such as the CBR circuits in ATM, are needed, this line must be configured as Fixed in the Behavior parameter. The foregoing means that if one needs to transmit services with voice or video, the ADSL line must be configured as Fixed.

4.2.4 QoS in MPLS

Two protocols have been defined in MPLS for QoS management. We will first discuss protocol the Constraint Routing–Label Distribution Protocol (CR-LDP) and then the Resource Reservation Protocol for Traffic Engineering (RSVP-TE), which specifies the same traditional RSVP protocol but applied in MPLS. For both cases

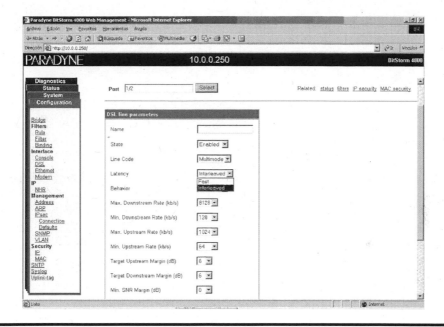

Figure 4.15   QoS parameters in ADSL.

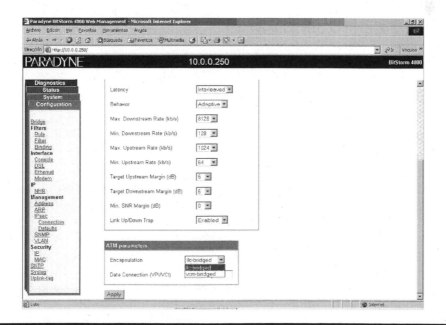

Figure 4.16   QoS parameters in ADSL.

the use of explicit routes (ERs) has been defined; this specifies the path from the source LSR to the destination LSR.

4.2.4.1 CR-LDP

CR-LDP is a label distribution protocol designed specifically to support quality of service. This protocol allows the establishment of explicit routes, with quality of service attributes on each segment of the defined route; it also allows reservation of resources [ASH02; JAM02].

This protocol is a set of LDP extensions specifically designed to facilitate routing based on LSP restrictions. LDP is the protocol that allows an LSR to inform another of the different labeling assignments that it has made. CR-LDP is used to distribute labels and establish and maintain LSPs along the explicit route or the constrained route. LDP assigns a FEC to every LSP and the LSPs spread through the network as every LSR assigns an ingress label for an FEC. There are four categories of LDP messages:

DISCOVERY Messages—used to announce and maintain presence of an LSR in a network and to enable LSP peers to establish label distribution connections

SESSION Messages—used to establish, maintain, and conclude sessions between LSP peers

ADVERTISEMENT Messages—used to create, change, and erase label mapping messages for FECs

NOTIFICATION Messages—provide administrative and error signaling

The new characteristics are explicit routing, resource and class reservation, LSP Identifier (LSP ID), Path Preemption, Route Pinning, and Failure Control.

Explicit Routing. As has been previously explained, explicit routing is when an LSR, generally the LSP ingress node or egress node, determines the complete route for the LSP. In CR-LDP an explicit route can also be called a *constraint-based route* or CR-LSP.

Resource and Class Reservation. CR-LDP allows resource reservation by explicit routes. The characteristics of the pathway can be described in terms of *Peak Data Rate* (PDR), *Committed Data Rate* (CDR), *Peak Burst Size* (PBS), *Committed Burst Size* (CBS), and *weight*. The PDR and CDR describe the bandwidth restrictions in the route. The PBS and CBS determine the maximum sizes of the rates and the weight determines priorities relative to every node. Later on we will explain the main functions of each of these parameters. There is also an option to indicate whether the resource requirement can be negotiable. If a parameter is negotiable, an LSR could specify a lower value for that specific parameter if it cannot be satisfied with the existing resources.

LSP Identifier (LSP ID). Every LSP is identified by an LDP IP, which is unique in the MPLS network. The identifier consists of the ingress LSR identifier and by a unique identifier for that LSR of the CR-LSP.

Path Preemption. If an LSP requires a certain reservation of resources and not enough resources are available, the new LSP may take away resources from the existing LSPs provided the new LSP has a higher priority than the existing ones. Two parameters are associated with an LSP for this purpose: the *setup* priority and the *holding* priority. The setup and holding priorities show the preferences to add a new LSP and retain an existing LSP. A new LSP can preempt an existing LSP if the setup priority of the new LSP is higher than the holding priority of the existing LSP. These priorities have values in a range from 0 to 7, where 0 is the priority assigned to the most important routers or the highest priority routes.

Route Pinning. CR-LDP can reoptimize an LSP. An LSP ID can be used to allow double reservation during optimization. Under certain circumstances route changes are not a good option. CR-LDP has a route pinning option. When this option is used, an LSP cannot change its route once it is established.

Failure Control. CR-LDP has a mechanism to detect signal communication failures. A recovery procedure must guarantee integrity of the connection at both ends of the channel. CR-LDP provides varied mechanisms for failure control, such as rerouting by means of modifications to the LSPs and fast rerouting by means of backup segments.

The CR-LDP protocol contains the following fields associated to the QoS parameters. Some of them may also be clearly identified in ATM networks or in Frame Relay networks.

- Flags—Identifies which of the following parameters are available for the negotiation:
  - Peak Data Rate (PDR)
  - Peak Burst Size (PBS)
  - Committed Data Rate (CDR)
  - Committed Burst Size (CBS)
  - Excess Burst Size (EBS)
  - Weight
  - Only the network nodes can change the parameter values. One node may overwrite the values of the parameters changed before resending the message to the next node and a node can change its reservation level based on the new parameters. This field is 1 byte long.
- Frequency—Specifies how frequently the data transmission average date (CDR) is ensured in the CR-LSP. This field is 1 byte long. Possible values:
  - Very Frequently—Indicates the number of delays required in the service. If the services do not tolerate long delays, they must be configured with

a Very Frequently value. Using a Very Frequently value one ensures the availability of transmission resources in an LSP for a transmission rate. This value is used for applications such as video conference.

- Frequently—Indicates that the applications can tolerate delay variations. This value is used for applications such as e-mail.
- Unspecified—Indicates that applications can be guaranteed over a long period of time. This value is used for applications that require better effort, such as Internet access.

■ Weight—Determines the relative priority for multiple LSPs when there is congestion or little bandwidth available.

■ Peak Data Rate (PDR)—Determines the maximum value of a transmission rate with which traffic should be sent via the CR-LSP.

■ Peak Burst Size (PBS)—Establishes the maximum amount of information to transmit at PDR.

■ Committed Data Rate (CDR)—Determines the average information transmission rate that an MPLS domain confirms available for the CR-LSP.

■ Committed Burst Size (CBS)—Establishes the maximum amount of information to transmit at a CDR rate.

■ Excess Burst Rate (EBR)—the amount of information to transmit by excess at a CDR rate.

The following are the different types of LDP messages with an emphasis on messages used in CR-LDP.

4.2.4.1.1 LABEL REQUEST Message

An LSR will send a LABEL_REQUEST to establish the labels necessary for transmission of an FEC through an LSP.

■ Message ID—32-bit value used to identify this message.
■ FEC TLV—The FEC for which the label is requested.
■ Optional Parameters—This variable length field contains 0 or more parameters, each one codified as a TLV. Optional parameters are
    - Hop Count—specifies the number of LSR hops along the LSP that are being established by the LABEL REQUEST message.
    Path Vector—Specifies the LSRs along the LSP that are being established by the LABEL REQUEST message.

The REQUEST message is used by an upstream LSR to explicitly request that the downstream LSR assign and publish a label for a FEC. An LSR could transmit a REQUEST message under any of the following conditions:

- The LSR recognizes a new FEC by means of the delivery table, and the next hop is a peer LDP and the LSR does not yet have a mapping from the next hop for the given FEC.
- The next hop for the FEC changes and the LSR does not yet have a mapping from the next hop for the given FEC.
- The LSR receives a LABEL REQUEST for an FEC from a peer LDP that is on the higher link, and the LSR does not yet have a mapping of the next hop.

The receiving LSR should respond to a LABEL REQUEST message with a LABEL MAPPING message with the requested label or with a NOTIFICATION message indicating that the request cannot be met.

The LABEL REQUEST message identifier serves as an identifier for the LABEL REQUEST transaction.

When the LSR receiver responds with a LABEL MAPPING message, the MAPPING message must include a LABEL REQUEST/RETURN message, which will in turn include the ID of the LABEL REQUEST message.

4.2.4.1.2 LABEL MAPPING Message

An LSR sends a LABEL MAPPING message to a peer LDP to publish the associated FEC. The coding of the message is as follows:

- Message ID—32-bit value used to identify this message.
- FEC TLV—The FEC for which the label is requested.
- Optional Parameters—This variable length field contains 0 or more parameters, each coded as a TLV.

A MAPPING message is used by an LSR to distribute a label mapping for an FEC to a peer LDP. An LSR is responsible for the consistency of the mapped labels it has distributed.

4.2.4.1.3 CR-LDP Functioning

- Ingress node LSR needs to establish a new LSP toward the destination LSR. The traffic parameters required for the session or the network's administration policies allow the source LST to determine the route that the new LSP should follow through the LSRs in the route to the destination LSR. The source LSR constructs a LABEL_REQUEST message with an explicit route and with the traffic parameters required for the new route. The source LSR reserves the resources necessary for the new LSP and then resends the LABEL_REQUEST to the next LSR.
- The LSR in the route receives the LABEL_REQUEST message, determines that it is not the egress node for this LSP, and resends the petition through

the route specified in the message. The resources required for the new LSP are reserved, the explicit route in the LABEL_REQUEST message is modified, and the message passes to the next LSR. If necessary, the LSR in the route may reduce the reservation for the new LSP, if the appropriate parameters were marked as negotiable in the LABEL_REQUEST.

■ The destination LSR determines that it is the egress node for this new LSP. Any final negotiation of resources takes place and the reservation is made for the LSP. A label is placed on the new LSP and is distributed to the previous LSR in the route with a LABEL_MAPPING message, which contains details of the final traffic parameters reserved for the LSP.

■ The previous LSR on the route receives the LABEL_MAPPING and places it along the original request using the LSP ID contained both in the LABEL_REQUEST message and the LABEL_MAPPING. The reservation is finalized, a label is placed on the LSP configuring the inlets of the routing table, and the new label passes to the source LSR in a LABEL_MAPPING.

■ The process at the source LSR is similar, but there is no need to place a label and resend it to a previous LSR because this is the ingress LSR for the new LSP.

4.2.4.1.4 Failure Detection and Error Report

CR-LDP uses HELLO LDP messages to detect neighbors and failures. DISCOVERY or HELLO messages are periodically sent as a package with UDP transportation protocol to the LDP pot of "all routers in the subnetwork" with group multicast address. A timer is associated to HELLO messages. There is also a keepalive timer for every session. If the timer expires, closing the transportation connection, it closes the LDP session and failure in the peer is assumed [JAM02].

Establishment of a CR-LSP using CR-LDP may fail for several reasons. All these failures are reported using the message LDP NOTIFICATION. This message generally carries a TLV status for reporting error reasons and codes.

4.2.4.1.5 Modifying LSPs Using CR-LDP

The purpose of this procedure is to change the resources reserved for an established LSP. It functions by creating a new LSP with the new resource requirements. Once the new LSP is created, the previous LSP is deleted, hence releasing the resources used by the previous LSP. Modifications are allowed only when the LSP is established and active. This means that modification is not defined or allowed while the LSP is being established or in the Release and Withdraw stages [ASH02b].

4.2.4.1.6 Rerouting LSPs

The purpose of rerouting LSPs is to provide a new route for a certain LSP due to notification failures, changes in topology, or changes in optimization purposes. It consists of creating a new LSP using at certain points of the path some LSRs different from the previous path. The mechanism supporting this function is known as *make-before-break* and consists of establishing a new path [ASH02b] before releasing the previous LSP.

4.2.4.1.7 Managing Priorities

LSPs may be differentiated by assigning them different priorities. When a LABEL_ REQUEST message is sent for an active LSP L1 with the purpose of requesting changes, the priority used in the LABEL_REQUEST message may be different from the one used in the previous LABEL_REQUEST message, effectively indicating the priority of this modification requirement.

Network operators can use this characteristic to decide what priority to assign to the modification requirement, based on its algorithms or policies or other network traffic situations. For example, the priority of a modification can be determined by the LSP client. If a client overly exceeds the reserved bandwidth reserved for its LSP Virtual Private Network (VPN) tunnel, priority of the modification requirement must be given as a higher value [ASH02b].

4.2.4.1.8 Modification Failures Management

A modification may fail due to insufficient resources or other circumstances. A NOTIFICATION message is sent back to the ingress LSR to indicate failure of the LABEL_REQUEST message that is trying to modify the LSP. If the LSP in the original route fails when a modification attempt is in progress, the attempt must be aborted using the LABEL_ABORT_REQUEST message. In case of a failure modification, all modification to the LSP, including the holding priority, must be restored to the original value [ASH02b].

4.2.4.1.9 Fast Rerouting

Establishment of backup LSPs has been specified in the CR-LDP so that in case of failures of the active LSP, the information transmission restoring mechanism can act as quickly as possible. These backups exist throughout the LSP sections, selectively protecting them from failures in the nodes and links. Backup segments have the provision to be shared among multiple LSPs across the protected set of links and nodes, so that a scalable solution is provided. We will next discuss exclusive bandwidth protection, shared bandwidth protection, and bandwidth protection without guarantees.

To satisfy real-time application needs, such as voice and video, in case of failure, it is necessary to reroute traffic to alternative routes within a few milliseconds. Since the local repair mechanism cannot compute and establish alternative routes in a short time period, the backup tunnels are used to carry traffic in times of failure.

When the backup segment is established and failure of a node or link occurs, traffic is commuted to the backup tunnel from the local repair point. Backup segments attempt to cover both node and link failures, until the local repair mechanism computes an alternative route for the primary LSP.

Two additional TLVs, TUNNEL_PROTECT TLV and BACKUP TLV, are proposed for the CR-LDP. The TUNNEL_PROTECT TLV will be used in the LABEL_REQUEST message while the BACKUP TLV will be used in the LABEL_REQUEST message and the LABEL_RELEASE message [ASH02b].

> TLV TUNNEL_PROTECT—Tunnel Protect is used by the ingress LSR to indicate that the tunnel has to be protected while a local repair is in progress. The backup may be selected without bandwidth protection, with shared bandwidth protection, and with exclusive bandwidth protection. Using shared bandwidth protection the bandwidth can be shared through the backup segment routes. *Without bandwidth protection* indicates that the backup segments do not need to have bandwidth reservation. *Without bandwidth protection* and *shared bandwidth protection* may be selected when no bandwidth assurances are necessary for the short period of time while the local repair is taking place.
>
> TLV BACKUP—The TLV backup is used throughout the tunnel backup session to signal that this is a backup session.

4.2.4.2 RSVP-TE

RSVP was initially designed to reserve resources in IP networks. RSVP-TE is an extension of the original RSVP protocol designed to distribute labels over MPLS; in addition, it supports the creation of explicit routes, with or without resource reservations. One of the most important additional characteristics of this protocol is that it allows rerouting of LSP tunnels, with the purpose of providing a solution for service outages, congestion, and bottlenecks [AWD01; WAN01]. The new characteristics added to the original RSVP include the following:

■ Label Distribution. Routers and hosts that support RSVP and MPLS may associate labels to the RSVP flows. MPLS and RSVP combine, making definition of a flow more flexible. Once the LSP has been specified, traffic through the route is defined by the label applied to the ingress node. When the traffic is mapped by means of labels it is possible at this time for a router to identify the appropriate reservation status of a package, based on the value of its label. The main difference between routes reserved by the original RSVP and the LSPs established by RSVP-TE is that in the former, the reserved route is associated with a specific destination and a transportation level protocol (TCP/UDP); the intermediate nodes will resend packets based on the IP header. With LSPs (RSVP-TE), the ingress node may determine which packets are sent through the LSP and which packets are sent to the intermediate nodes.

- ▪ Node Abstraction Concept. An abstract node is a set of nodes whose internal topology is transparent to the LSP ingress node. An abstract node is simple if it contains a single physical node. Using this abstraction concept, an LSP routed explicitly may be specified as a sequence of IP prefixes or a sequence of autonomous systems.

- ▪ Explicit Routing. An explicit route is a sequence of hops from the ingress node to the egress node. An LSP in MPLS can be configured to follow an explicit route. For example, a list of IP addresses could be an explicit route. However, it is not necessary to completely detail the route to be followed; the route could be specified only in the first hops. An explicit route may be classified as *strict* or *loose*. A strict route may contain only the nodes specified in the ER and must use them in the designated order, whereas a loose route must include all the specified hops and maintain order, and may include additional hops if necessary to reach the specified hops. An explicit route is useful because it can be used to distribute traffic in networks with high data-transmission levels or to assign backup LSPs in order to protect the network against failures.

- ▪ Bandwidth Reservation for LSPs. RSVP-TE has the option of reserving bandwidth for LSPs by means of the message PATH. The message RESV will return the label to be used.

- ▪ Rerouting of LSPs After Failures. It may be desirable to reroute LSPs under certain circumstances. For example, an LSP tunnel may be rerouted to restore connectivity after network failure. To implement rerouting RSVP-TE uses two techniques known as *optimized rerouting* and *fast rerouting*.

- ▪ Location of the Current Route of an LSP. The current route of an LSP must be periodically updated by the object RECORD_ROUTE.

- ▪ LSP Priority (Preemption Option). Preemption is when an LSP's resources are required for a new, higher priority LSP. The preemption option is implemented by two priorities: The setup priority serves to take resources and the holding priority is the one used to maintain resources.

4.2.4.2.1 Functional Specification

New objects were added to RSVP-TE in comparison with the RSVP traditional with the purpose of supporting traffic engineering [AWD01] (see Table 4.5).

All the new objects are OPTIONAL for RSVP; however, the LABEL_REQUEST and LABEL objects are mandatory for RSVP-TE.

Table 4.5

Object Name	Applicability in RSVP Messages
LABEL_REQUEST	PATH
LABEL	RESV
EXPLICIT_ROUTE	PATH
RECORD_ROUTE	PATH, RESV
SESSION_ATTRIBUTE	PATH

4.2.4.2.2 PATH Messages Format

The PATH message is responsible for establishing the status and requesting label assignments. The following shows the format of PATH messages for RSVP-TE.

4.2.4.2.3 RESV Messages Format

The RESV message is responsible for distributing labels and reservation of resources.

4.2.4.2.4 RSVP-TE Functioning

RSVP-TE is an adaptation of the original RSVP to support LSP tunnels and also to distribute MPLS labels [AWD01]. Next, we will explain the process to configure and establish an LSP tunnel.

■ The source LST determines the need to establish a new LSP whose egress node must be the destination LSR. By means of the administrative policies, the source LSR determines the route for the new LSP; that is, it decides that the new LSP should be established through the intermediate nodes. The source LSR is responsible for constructing the PATH message with an explicit route, indicating the intermediate LSR on the route toward the destination LSR and with details of the traffic parameters requested for the new route. The source LSR sends the PATH message as an IP packet to the next LSR node specified as the next hop in the explicit route.

■ The intermediate LSR node receives the PATH message. It determines that it is not the egress node for this LSP. Then, it resends the request along the specified route by means of the EXPLICIT_ROUTE object, in this case, the destination LSR node.

- The destination LSR node determines that it is the egress node for the new LSP. It assigns the requested resources and selects a label for the new path. This label is distributed by means of the LABEL object through the RESV message.
- The intermediate node receives the RESV message of the destination LSR. With the parameters of this message it determines what resources to reserve, assigns a label for the LSP, and updates its routing table. Finally, it sends the RESV message to the node indicated, which, in this case is the ingress node.
- Processing at the LSR source node is similar; it just is not sent or assigned a new label because it determines it is the LSP ingress node. The LSP tunnel is now configured and established.

4.2.4.2.5 Failure Detection and Error Reporting

In RSVP-TE, PATH and RESV message update allows discovery of the active links. However, this is not considered a robust way of failure detection because the updating counter will have to be configured with a very small value for it to be moderately effective. To alleviate the resulting disadvantages and maintain robust failure detection, the message HELLO is used. This mechanism allows node-to-node failure detection. Another way of rapidly detecting failures is by means of the objects NOTIFY and NOTIFY REQUEST [AWD01].

4.2.4.2.6 Rerouting

Network topologies are unstable through time. For this reason, if a traffic engineering system wishes to provide good, optimized service, it must have the capability of responding to change. This is accomplished through rerouting [AWD01]. An LSP tunnel must be rerouted for the following reasons:

- Link failure. This rerouting is known as *Local Recovery*.
- A better route is available.
- An LSP's resources are required for a new LSP, which has a higher priority. When rerouted for this reason, this is known as *Preemption*.

There are two types of rerouting:

- Fast rerouting—used to minimize traffic interruption when the network is out of service. Fast rerouting is implemented by means of temporary or backup LSP tunnels, so that local LSP tunnels may be repaired. There are two basic techniques to configure backup tunnels: one-to-one backup and facility backup. The one-to-one strategy operates according to a backup LSP for every protected LSP, while the other strategy focuses on using a single LSP as backup for a set of protected LSPs.

■ Optimized rerouting—used to optimize traffic flows in a changing technology. Whereas fast rerouting is useful because it allows traffic transfer from one route to another without interruption, optimized rerouting allows construction of an optimum route and transferring traffic without greater interruption problems.

4.3 QoS in IP (DiffServ)

DiffServ is a CoS model that uses Internet better efforts services: differentiate traffic by user, service requirements, and other criteria. Hence, it marks packets so that the network nodes can provide the different service levels, whether by queuing priorities or assignment of bandwidth, or by selecting dedicated routes for specific traffic flows.

Several techniques that attempt to provide predictable Internet services have been proposed and developed. One technique is integrated services (IntServ) and its associated protocol, RSVP. The traditional Internet best effort model does not attempt to differentiate between the traffic flows that are generated by the different hosts. As the traffic flow varies, the network provides its best service, but there is no control to maintain high service levels for certain flows and not for others. What DiffServ does is try to provide better service levels in a better effort environment. IntServ is a bandwidth reservation technique that constructs virtual circuits through the Internet, where the bandwidth requests originate in the applications being executed at the hosts. Once the bandwidth reservation has been made, it cannot be reassigned or pretended by another reservation or traffic. IntServ and RSVP are stateful, based on status; that is, the nodes of the RSVP network must coordinate with the others to establish an RSVP route and remember the information about the flow status. This may be a very exhaustive task in the Internet where there may be millions of flows through a router. This proposal is considered very complex for Internet, but appropriate for smaller corporate networks.

DiffServ is stateless, that is, not based on status, as it minimizes the number of nodes required in the network to remember the flows. DiffServ is not as good as IntServ in QoS, but its implementation is much more practical in Internet. DiffServ edge devices mark packets to describe the level of services they must receive. The network elements simply respond to such markings without having to negotiate routes or remember large amounts of information for every flow. Moreover, applications do not need to request a specific service level or provide greater information about the direction of the traffic. In the IntServ/RSVP environment, applications negotiate with the network for the service. IntServ is aware of the applications, thus allowing hosts to communicate useful information to the networks about their requirements and the status of their flow. In turn, DiffServ is not aware of the applications (although internally it works to do so). Since DiffServ does not listen to the applications, it does not benefit from the feedback that such applications could provide, and, for this reason, DiffServ does not know what the applications really require. Therefore, there is a possibility of failure to provide the appropriate service level. Furthermore, DiffServ is not in contact with the other receiving hosts and, therefore, does not know whether a host can manage the services it will provide. One could say that Internet needs both technologies: RSVP (or any other complete QoS model) and DiffServ. The idea is to combine IntServ and DiffServ into an end-to-end model, with IntServ as the architecture that allows applications to interact in the network and DiffServ as the architecture to manage network resources. We have so far compared DiffServ with IntServ and RSVP, but DiffServ could also be contrasted with MPLS, which implements virtual circuits aimed at ATM, Frame Relay, or commuted network connectivity. MPLS marks packets indicating reforwarding behavior, but they travel through predefined circuits. MPLS is generally more sophisticated and complex than DiffServ, but provides better QoS capabilities.

A differentiated services architecture must maintain a simple reforwarding route, transfer network complexity as closest to the edge as possible, provide a service that avoids assumptions about the type of traffic using it, use an assignment policy that is compatible with the short-term and long-term requirements, and maintain the best effort model for the dominant Internet traffics. Some important aspects about DiffServ include the following:

- DiffServ defines a new field, Differentiated Service (DS) in the IP header that replaces the old Type of Service (ToS) field. The bits sequence in the field indicates the type of service and reforwarding behavior in the network nodes.
- The 6 bits in the DSCP field may define up to 64 types of network service.
- Per Hop Behavior (PHB) refers to a specific packet reforwarding treatment through the DiffServ network. The value in the DSCP field indicates the PHB to be used.
- PHBs may satisfy specific bandwidth requirements (i.e., real-time voice support) or provide some priority service. The characteristics of the service may be designed to improve forwarding or reduce delays and lost packets.
- Differentiated services may accommodate different types of application requirements and allow service providers to set prices according to their characteristics.

Operations consist of traffic classification. This is done by awarding specific treatment to every packet based on the mark in the ToS field in the IP header, which is now called DS CodePoint. All the packets having the same mark are treated equally.

Segments are marked anywhere in the network (although generally they are done at an end of a domain), and no extra flow with information on the treatment to be awarded each segment is needed; rather, the task to be optimized goes in the ToS field.

DiffServ services are based on rules to define the way in which segments will be initially marked and the manner in which they will be treated by other nodes. In addition, a bilateral agreement between the points to communicate is necessary.

The three elements that work to provide DiffServ services are as follows:

- Per Hop Behavior (Behavior between hops mechanisms, PHB)—a set of rules or policies that define treatment of packets when being marked
- Traffic Conditioners—alterations added to packets to force compliance of the service rules
- Bandwidth Brokers (BBs)—manage communications policies

The ToS field also plays an important role in DiffServ's operation. In this QoS scheme it takes the name DiffServ codepoint (DSCP), in which the last two bits are reserved and six are used for the service code, thus allowing differentiation of 64 services, 48 for global use and 16 for local use.

The architecture of differentiated services is based on the classification and conditioning of incoming traffic at the edge nodes, where they are provided a behavior aggregate, BA. Every BA is identified by its DSCP. Within the network, packets are treated according to the PHB associated to their DS codepoint.

DiffServ architecture may be analyzed under different domains.

DiffServ Domain (DS Domain)—A DS domain is a set of nodes that operate with a set of common services, PHB groups, and policies applied to all nodes. A domain consists of boundary nodes and interior nodes. Interior nodes manage only traffic within the networks depending on the PHBs, and the boundary nodes ensure classification of the traffic according to this DSCP and the interconnection with other DS or non-DS domains. Nodes or equipment on a DS domain must have DiffServ accessibility or capability, such as the inclusion of points without DiffServ application capability; that is, not being able to apply necessary packet behaviors might produce unexpected effects on the traffic.

Differentiated Service Region (DSR)—A DSR is a set of one or more contiguous DS domains. DS regions are capable of supporting differentiated services along the paths that span the domains within the region. The regions may support different PHB groups, that is, different correspondences of DSCP ↔ PIIB. To control communication across the different domains in the region, a TCA and an SLA must be established; or, better yet, one of the same PHB groups should be adopted, that is, the same DSCP ↔ PHB correspondence.

The most important parts for operation of the DiffServ architecture are PHB, classifiers, conditioners, and bandwidth brokers.

4.3.1 PHB

PHB is the definition of the treatment given to packets based on the marking policies; that is, it is the one carrying the DS to the packet.

IETF currently defines three PHB classes: expedited forwarding (EF), assured forwarding (AF), and default (DE).

EF defines a sort of dedicated virtual line between the two nodes communicating, and it is required that the sending node have a larger rate of egresses than ingresses. In addition, a small buffer is necessary, as well as priority management, to manage queues. This PHB must be used only by a small traffic fragment in the network.

AF establishes four levels of traffic priority, which determine the bandwidth to assign to each type of packet, namely, gold, silver, bronze, and better effort. In this service scheme, packets with the lowest priority are most likely to be eliminated or not reach their destination. A regulating mechanism is required for each priority level and another one is needed to regulate bandwidth management and control congestion.

DS leaves the ToS byte open, restructuring it to the DSCP form for future incorporation of a PHB.

If packets with the DS byte span several DiffServ domains, they must each support the same PHB.

4.3.2 Classifiers

Traffic is classified taking into account DSCP according to the PHB it contains. DiffServ supports a maximum of 64 services. However, these services depend on the configuration made by the network administrator or the vendor of a product with built-in DiffServ, as different codes may be found for the same PHB. For example, the code "011010" may belong to the same PHB as "011000."

Classification is performed by a router that orders packets within the queues according to the information in their DS field. Every queue may have a different treatment according to priority, assigned bandwidth, or discarding policies.

4.3.3 Traffic Conditioners

The purpose of traffic conditioning is to ensure that traffic entering the domain conforms to the rules in accordance with the domain's service provisioning policies.

Every router is enabled to provide traffic conditioning services by metering, controlling, guiding, and marking packages to ensure that traffic managed in the network with service differentiation is managed according to the traffic conditioning agreements, TCAs. This process is described in the following paragraph.

All traffic-related agreements are found within an agreement level called Service Level Agreement (SLA).

The following are the functions performed by the conditioning components:

- Monitoring. Monitoring traffic patterns according to their traffic profile. For traffic without profiles, a joint decision is made with other parties to re-mark or discard the packets.
- Marking. Every service requires a value in DSCP according to the level of service required by each package to define the PHB that is stored in the packet. This is done so that edge routers can classify the packets according to their PHB and their DSCP.
- Guide/Control. When a packet enters a router, the router classifies it according to the packet's PCB, but if it does not have a service profile, it is discarded or re-marked. Also, it regulates the traffic established for every type or profile of service, thus ensuring justice for the different classes of services, and controls overall traffic to prevent congestion.

4.3.4 Bandwidth Brokers (BBs)

BB was designed according to network organization policies to manage location of marked traffic. Its function is to guarantee a bandwidth through all the network elements in an end-to-end connection, according to the required service.

Multilateral agreements for DiffServ domains are infrequent because of the communication needs of a bilateral agreement between the two points to communicate: the BB needs only to establish relations with a certain level of trust between the communicating nodes.

DiffServ architecture offers a range of services for each network that are monitored with justice and with agreements between consumers and service providers.

Figure 4.17 shows how QoS can be configured in DiffServ in an IP-PBX for transmission of IP telephone calls with a DSCP value of EF, that is, the better quality that can be specified. Figure 4.18 shows how to configure DiffServ (DSCP field) in a LAN switch, and Figure 4.19 shows how one can configure a VoIP flow in a DiffServ bandwidth manager.

4.4 QoS in Layer 4 (TCP/UDP Port)

An alternative way to apply QoS, but in this case associated to the application, would be through layer 4, that is, through the TCP or UDP port being used by the application. For this case, the equipment that must understand the value of port TCP or UDP must be layer 4. This function may be found in most LAN switches or routers. The objective in this case is for the equipment that is going to apply QoS to review and understand the value of the port, whether TCP or UDP, and to subsequently apply the QoS parameters specified for this specific specified flow or application.

Figure 4.20 shows an HTTP packet through a sniffer, where one can see that the destination port field of header TCP in this case indicates a value of 8080, which represents the standard port of Proxy HTTP. In conclusion, if the equipment

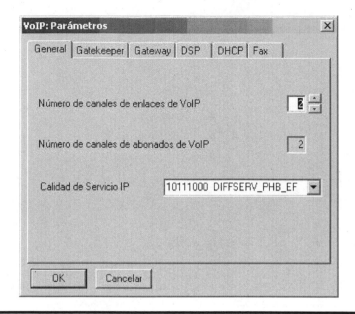

Figure 4.17   DiffServ configuration in an IP-PBX.

is capable of understanding the TCP port, it will apply the QoS parameters according to the procedure previously explained.

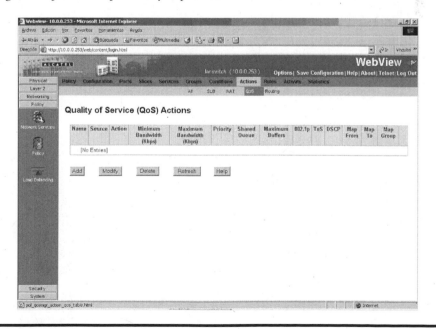

Figure 4.18   DiffServ configuration in a switch LAN.

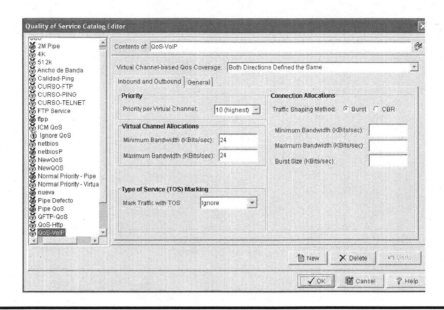

Figure 4.19   DiffServ configuration in a bandwidth manager.

Figure 4.21 shows a File Transfer Protocol (FTP) packet. Here, one sees that the destination port value is 21, by which this application can be identified and subsequently apply the QoS parameters.

Figure 4.20   TCP port in an HTTP application.

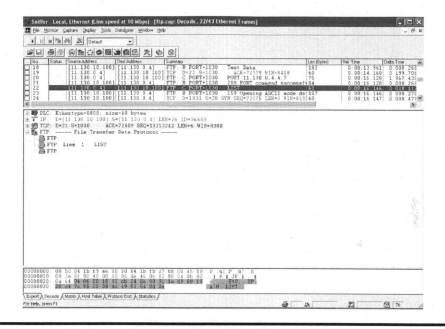

Figure 4.21   TCP port in an FTP application.

In the case of voice or video transmission, these will be transmitted via UDP packets. Figure 4.22 shows a VoIP packet in which a UDP with a value of 49604 is used.

From the examples shown, we can realize that when QoS is applied in layer 4 it is possible to understand up to ports TCP or UDP, but it will not be possible to understand what application is exactly being transmitted when, for example, the TCP or UDP ports are variable and this situation occurs in VoIP, Videoconference over IP, and IPTV transmissions.

Figure 4.23 shows how a layer 4 device can understand the TCP or UDP port and hence identify a specific flow or application.

Once the flow or application through port TCP or UDP is identified, one must define the QoS parameters to be applied. Figure 4.24 shows some QoS configuration parameters in an OmniSwitch Alcatel 7700. As seen in the figure, it is possible to configure minimum and maximum bandwidth; priority, which is directly associated with delay; jitter, if the queue is shared or dedicated; and the buffer space, and jointly it can be marked with the field 802.1p in Ethernet or with the DiffServ field DSCP in IP.

Hence, once the application has been identified, such defined QoS parameters would be applied.

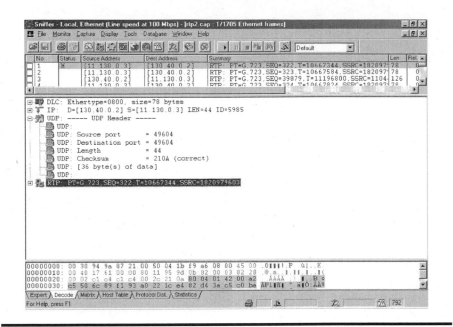

Figure 4.22   UDP port in a VoIP application.

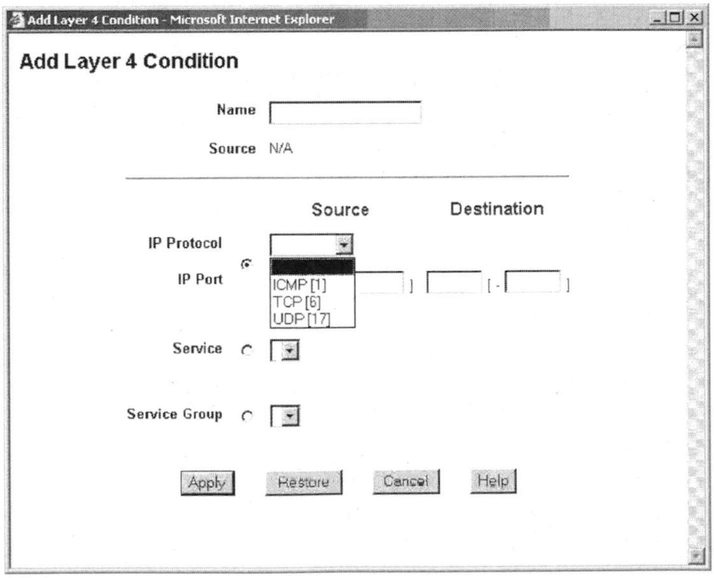

Figure 4.23   TCP or UDP port in a switch.

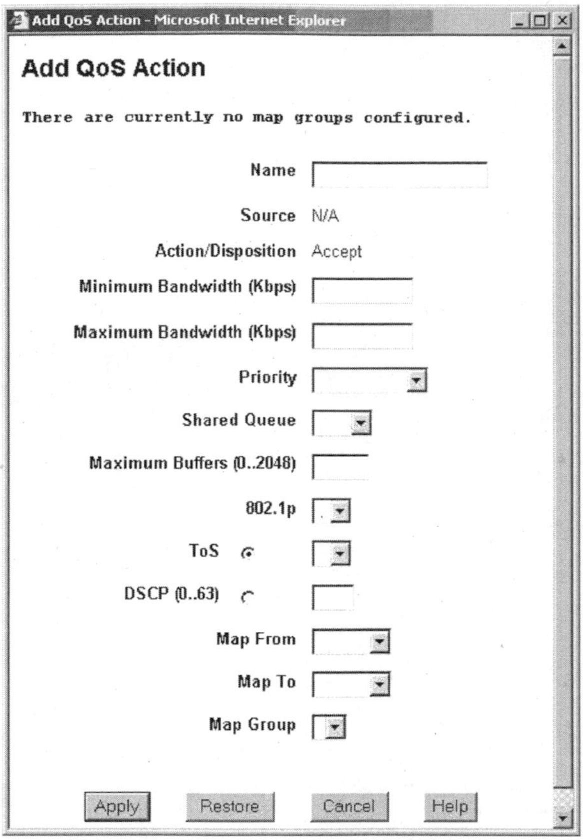
Add QoS Action - Microsoft Internet Explorer
Add QoS Action
There are currently no map groups configured.
Name
Source N/A
Action/Disposition Accept
Minimum Bandwidth (Kbps)
Maximum Bandwidth (Kbps)
Priority
Shared Queue
Maximum Buffers (0..2048)
802.1p
ToS
DSCP (0..63)
Map From
Map To
Map Group
Apply
Restore
Cancel
Help

Figure 4.24 QoS parameters.

4.5 QoS in Layer 7 (Application)

Another way of identifying the flow to be transmitted with QoS is analyzing the specific application being transmitted notwithstanding the TCP or UDP ports being used. The typical case, or the one generating the most inconveniences to be analyzed through layer 4, is voice or video transmission, because the ports used in service transmission are variable and normally depend on the manufacturer of the equipment over which the service is taking place. Exceptions to this are the control and signaling messages, which are indeed standard and are associated to either an H.323 or an SIP. A signaling message, for example, is the one used by port TCP 1720.

In order to analyze which application is being transmitted, one needs equipment traditionally denominated layer 7, because this is the application level in the OSI model. There may be layer 7 switches on the market, but the most common layer 7 equipment are the bandwidth managers such as Packeteer or Netenforcer.

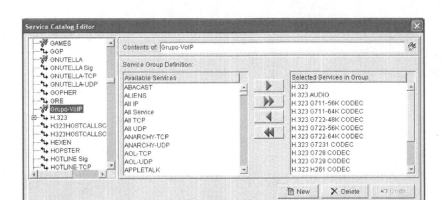

Figure 4.25   VoIP application.

Figure 4.25 shows how a Netenforcer understands a specific application such as VoIP notwithstanding the codec being used or if port TCP 1720 is used to signal between two other types of packets and ports. A group called *Group-VoIP* has been defined in this case, wherein are defined the set of H.323 standards, different codecs, and also the signaling messages.

With layer 7 equipment one can define the different services over which one wishes to apply QoS. Figure 4.26 shows a sample configuration performed with the

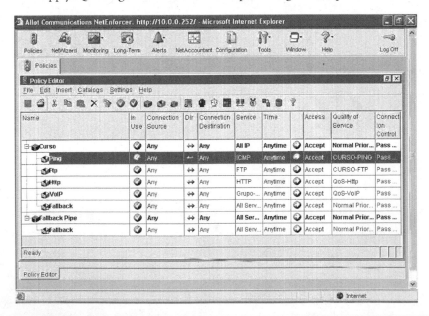

Figure 4.26   QoS applications.

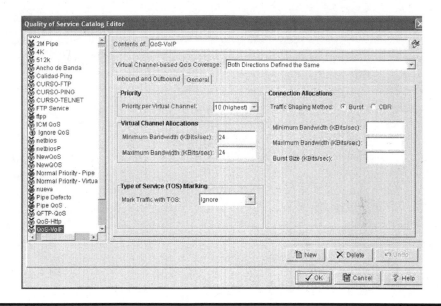

Figure 4.27   VoIP parameters.

same bandwidth manager, Allot Netenforcer. Here, one can see that four different services have been defined: Ping, FTP, HTTP, VoIP, and Fallback for the rest of the applications to which the best effort would be applied. For every type of service a different management for QoS parameters would be specified.

Figure 4.27 shows an example of certain QoS parameters that have been defined for the VoIP service. This example shows a constant value of 24 Kbps reserved for the VoIP service through the minimum bandwidth and maximum bandwidth parameters; in addition, it has been assigned the highest priority with the value 10.

4.6 Network Design with Bandwidth Manager

Through the different sections of this chapter we have seen that QoS characteristics can be configured in different types of equipment such as switches and routers, and QoS policies can be configured through different schemes such as physical ports, VLANs, IP routing, TCP or UDP ports, and applications. The foregoing suggests that the QoS concept may even be applied on extreme equipment in transmission of the service.

Even though QoS may be configured through different equipment, there are some called *bandwidth managers* that have been designed and manufactured specifically for defining and applying QoS. This section will describe where to place the bandwidth manager in a network design.

Figures 4.28 and 4.29 show that the traditional location of the bandwidth manager is between the main core switch of the LAN network and the connecting router toward the MAN or WAN in the last mile to the carrier. This shows that when a QoS application in critical services is required, the bandwidth manager will be responsible for performing this function in the edge toward the connection with the carrier where the lowest transmission rates (in Mbps) compared with a LAN network where transfer rates are higher (in Gbps).

Figure 4.30 shows an example of a bandwidth manager device. In this case we show an Allot Netenforcer. The bandwidth manager connects to the main switch of the LAN network through the internal Ethernet port and connects to the router connecting to the MAN or WAN through the external Ethernet port.

Once the QoS services and parameters have been configured in the bandwidth manager as shown in the previous QoS section in the application level, the services could be analyzed to verify that they are in compliance with the QoS policies applied. Figure 4.31 shows an example analyzing the bandwidth parameter for VoIP and FTP services. In this case, 24 Kbps have been reserved for VoIP service and 15 Kbps for FTP.

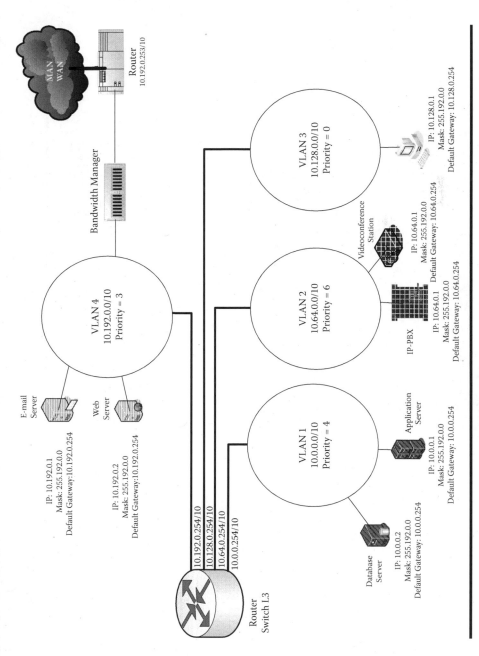

Figure 4.28 Logical scheme of the bandwidth manager.

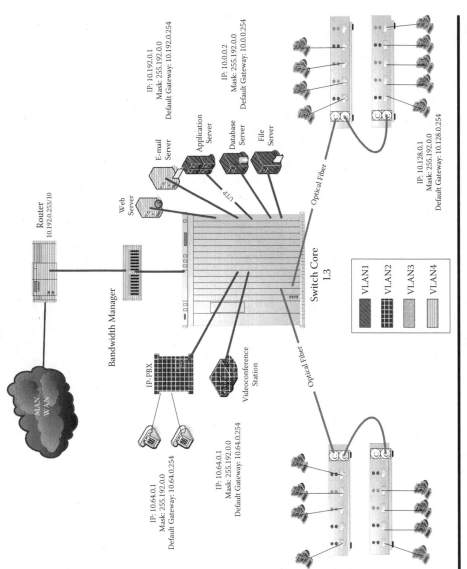

Figure 4.29 Physical scheme of the bandwidth manager.

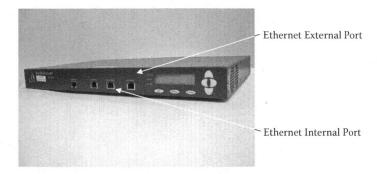

Figure 4.30 Bandwidth manager: Ethernet external port; Ethernet internal port.

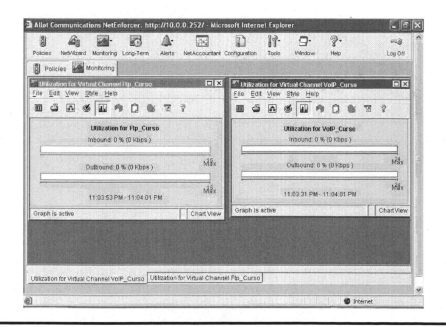

Figure 4.31 Bandwidth manager monitoring.

Chapter 5

Computer Network Applications

The purpose of this chapter is to integrate certain applications with the infrastructures of convergent networks that we have discussed thus far in previous chapters. In this chapter we will see that the type of service notwithstanding, voice and video can be transmitted over the same IP network. Due to the nature of the different types of applications it is necessary to specify certain QoS parameters, which were precisely discussed in the previous chapter.

We have divided this applications chapter in two parts. The first part covers not real-time applications; this means that the QoS requirements are not so critical. Typically, these applications correspond to data transmissions exclusively. The second part discusses applications commonly grouped as real time; this means that the QoS requirements are critical. Typically, these applications correspond to transmission of critical data or voice or video transmission. The purpose of this chapter is not to detail how these applications function, because there are other texts focused on that subject, but to show and integrate applications to the network infrastructures that we specified in the previous chapters.

5.1 Not Real-Time Applications

Because the service convergence concept involves data, voice, and video transmission, we must also discuss, in a general scheme, the traditional transmission of data applications. Although we might think that the most worrisome applications are those that transmit voice or video, it is also true that there are certain data applications that might become worrisome due to the expected values such as, for example, in response time that users have.

In this not real-time section, we will show some characteristics of certain traditional applications. It is not the intention of this book to be exhaustive with respect to the different applications that might be in operation, but to show how such applications integrate with the designs shown in previous chapters and to certain quality of service requirements that may be established.

5.1.1 HTTP

The first application that we are going to discuss is Hypertext Transfer Protocol (HTTP), which is one of the most common in Internet.

In order to transmit information one must have a server that contains, among different aspects, the site's Web pages. We say different *aspects* because the Web site might be connected to databases for the acquisition or registration of information, or it might be connected to security systems or authentication systems, among others. Therefore, the Web site becomes a robust applications scheme that works collaboratively. For this case, we will discuss it only from the perspective of data transmission via HTTP between a server and a client surfing the Web.

Figure 5.1 shows a connectivity scheme between a client application and a server in HTTP. The figure shows that a client requests a connection toward the HTTP server, and once the server receives this request, it will respond sending the content of the Web site. For this example we are assuming the existence of an IP network such as described in previous chapters. In fact, when applications work over TCP or UDP, this is the perspective they see of the transmission, regardless of how many routing or switching functions, among others, must take place for the information to travel between the server and the client.

Figure 5.2 shows a scheme contemplating different transmission levels and technologies to effectively perform transmission of such application. In this example we have IP over GMPLS. As to layer 3 in this example, transmission would take place between the client, whose IP address is 10.1.0.100, and the server, whose IP address

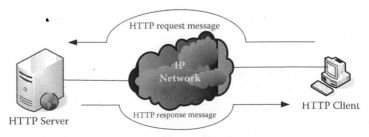

Figure 5.1   HTTP connection.

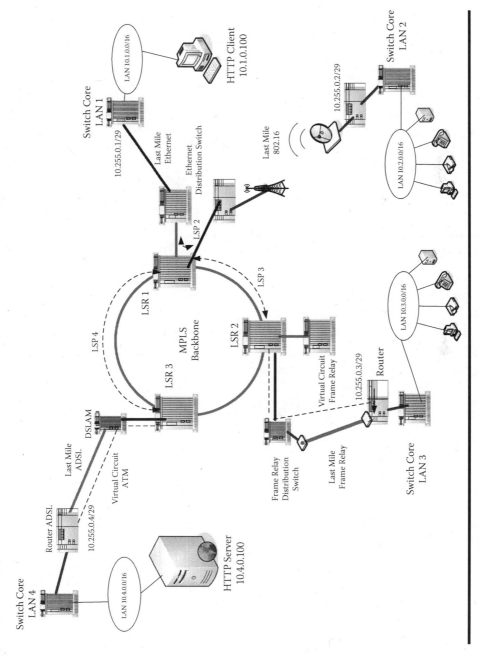

Figure 5.2 HTTP application over IP/GMPLS network.

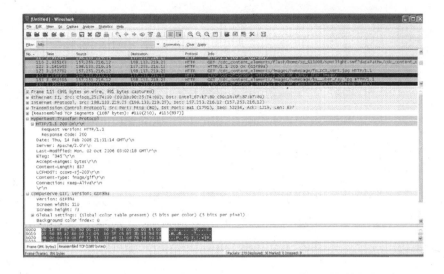

Figure 5.3   HTTP application.

is 10.4.0.100. Regarding layers 1 and 2, we have Ethernet in the LAN networks, with Ethernet in the client last mile, ATM/ADSL in the server last mile, and MPLS in the network operator backbone. We can appreciate the integration of the different technological platforms to provide solutions to the convergence of services in IP networks.

In Figure 5.3 we can see through the Wireshark sniffer an HTTP packet in which one sees the different levels. In this case we can see the HTTP application being transmitted over a TCP connection, and this connection, in turn, over IP.

5.1.2 FTP

Similarly, File Transfer Protocol (FTP) requires a server with which the client wishes to communicate, whether to forward the file toward the server or download it from it. The communication between the client and the server will take place through port TCP 21 and transmission of data through port TCP 20 (Figure 5.4).

Figure 5.5, as shown in HTTP transmission, shows for FTP the same network scheme contemplating different transmission levels and technologies. In this example we have IP over GMPLS. As to layer 3 in this example, transmission will take place between a client, whose IP address is 10.1.0.100, and the server, whose IP address is 10.4.0.100. As to layers 1 and 2, we have Ethernet with the LAN networks, Ethernet in the client last mile, ATM/ADSL in the server last mile, and MPLS in the network operator backbone. We can appreciate the integration of the different technological platforms to provide solutions to the convergence of services in IP networks.

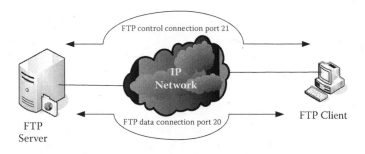

Figure 5.4 FTP connection.

In Figure 5.6 we can see through the shown Wireshark sniffer an FTP packet in which one sees the different levels. In this case we can see the FTP application being transmitted over a TCP connection, and this connection, in turn, over IP.

5.1.3 SMTP and (POP3/IMAP)

Another service associated to data transmission that we can mention is electronic mail. As is well known, through this service we can send information, files, images, etc., to different persons. The protocols used for mail transmission include Simple Mail Transfer Protocol (SMTP), Post Office Protocol–Version 3 (POP3), and Internet Mail Access Protocol (IMAP). The last two protocols, that is, POP3 and IMAP, are used to access mail accounts and retrieve the information.

In order to be able to send this information between users, one must follow certain steps that are shown in Figure 5.7. User 1 wants to send mail to User 2 (❶). Once User 1 has finished writing the mail, the mail is sent to the server in User 1's mail account using the SMTP protocol (❷). Once User 1's server has received the message, using SMTP protocol, it will forward the message through the IP network to User 2's mail server (❸). Once this mail has reached User 2's mail server through the IP data network through protocol SMPT, User 2 may consult or download the message received. Therefore, step (❹) is that User 2 may consult or download through protocols POP3 or IMAP and even via HTTP the mail that has arrived at his server. Finally, User 2 may read the mail through any of the e-mail clients (❺).

There are many differences between POP3 and IMAP, but the most common one is that through POP3 mail is downloaded in the local user machine, while in IMAP, the idea is to keep mail and the mail administration structure in the server. Figure 5.8 shows a POP3 packet.

Finally, the connectivity scheme of both e-mail servers and clients is the same as those previously shown for HTTP and FTP.

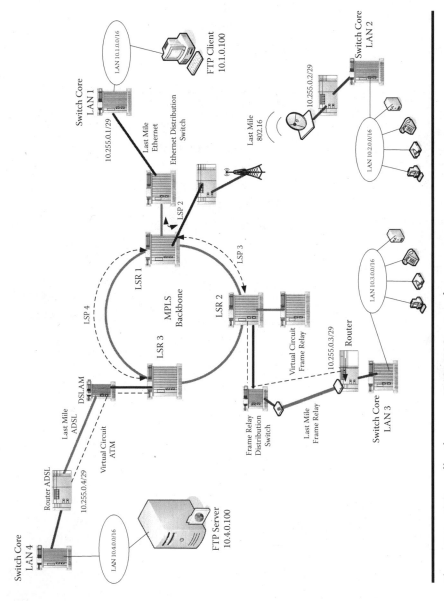

Figure 5.5 FTP application over IP/GMPLS network.

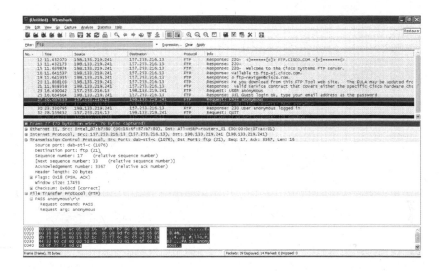

Figure 5.6 FTP application.

5.2 Real-Time Applications

Having explained the basic service scheme and connectivity to the IP network, we will continue with convergence of critical applications. In other words, transmission of voice and video over IP networks. Today, new networks have to be designed to support transmission of voice and even video services. These new networks, which are called *Next Generation Networks* (NGNs), have an architecture that integrates voice and video. Another term has been created, *IP Multimedia Subsystem* (IMS), for the architecture that delivers multimedia IP, including mobile networks.

In this section we will first discuss the services and next, each of the architectures that are integrated under a single IP platform.

5.2.1 VoIP

Voice over IP (VoIP) consists of transmitting voice signals through the IP data network instead of sending them as a digital or analog signal by means of a traditional Public

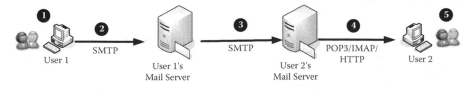

Figure 5.7 E-mail application.

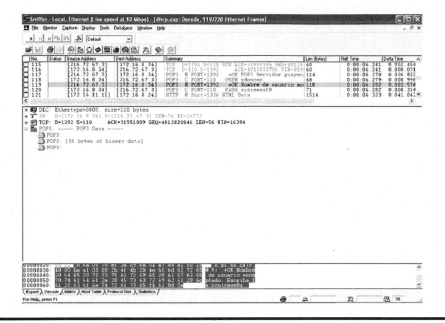

Figure 5.8 POP3 application.

Switching Telephone Network (PSTN). Voice signals are encapsulated in IP packets that can be carried as IP native or IP by Ethernet, Frame Relay, ATM, or SONET.

In practice there are two different ways of implementing voice over IP networks. In this section we will analyze two of them: through IP-PBX and under the IP cellular telephony scheme.

Figure 5.9 shows a VoIP transmission packet. As can be seen, codec G.723 is being used to digitalize the voice. This information is encapsulated within Real-Time Protocol (RTP). RTP, in turn, is within UDP and UDP is within IP. This is the protocol scheme that works voice or video, whether through H.323, Session IP Protocol (SIP), or other specified. This book does not discuss SIP or any other protocol designed for service convergence.

5.2.1.1 IP-PBX

Organizations may need to transmit different telephone calls over IP networks. This is most commonly case implemented through IP-PBXs. In this case, the IP PBX handles many telephone extensions that can be analog or digital and even IP telephones, which may be connected to the LAN network, whether by Ethernet or through WiFi. In turn, IP-PBX is one of the pieces of equipment responsible for redirecting the telephone call through the IP network.

Figure 5.10 shows a generic connectivity and voice transmission using IP-PBX as gatekeeper (H.323) or as media gateway controller or softswitch in SIP. In this

Figure 5.9   VoIP packet.

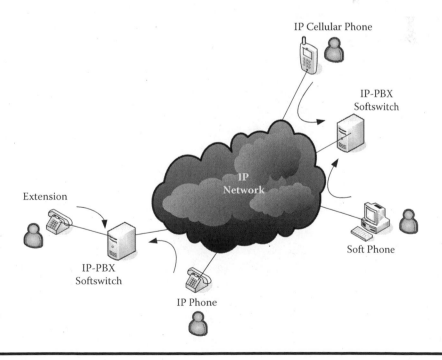

Figure 5.10   VoIP communication.

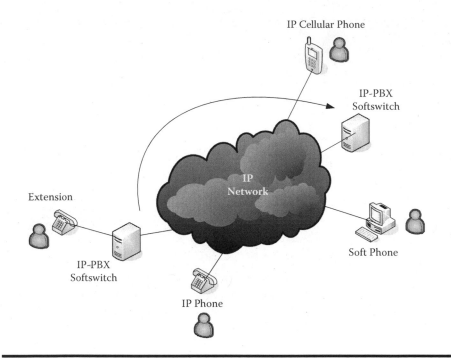

Figure 5.11 VoIP communication.

example, every voice device must establish connection toward the destination voice equipment through the softswitch, as can be seen in Figure 5.10. In this case, when the voice terminal equipment wants to establish a call, it connects to the softswitch.

Subsequently, the softswitches will establish connectivity between them in order to connect with the destination voice equipment. Figure 5.11 shows that there are messages between the softswitches to establish connectivity.

Next, the softswitch sends the connectivity message to the voice destination equipment to establish the telephone call. Figure 5.12 illustrates this scheme.

Once connectivity is established, the voice transmission will take place between the two end points, which in this case could be, for example, from a telephone extension through IP-PBX or an IP telephone to a SoftPhone. Figure 5.13 shows that the voice packet transfer takes place directly between the end equipment.

In addition to these voice packets, other control packets will be sent between the end equipment and the softswitches or among softswitches.

As we have seen in the previous figures, at this level of the application we now see an IP cloud as the network that will serve as connectivity platform for the applications, whether real time or not real time. But this IP cloud will be determined the same way as explained in previous chapters. For this reason it is necessary to define a good networking infrastructure with appropriate QoS parameters to ensure coexistence of data services as well as voice and video. The scope of this book does not

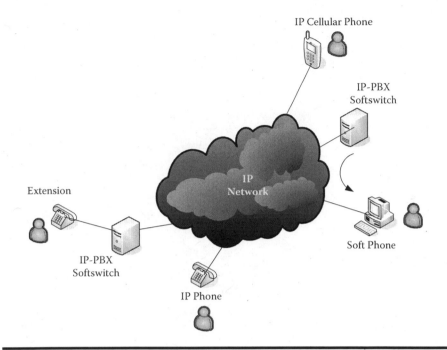

Figure 5.12   VoIP communication.

cover specification of protocols H.323 or SIP or any other messages, because there are many good books or documents whose purpose is to exactly specify them.

Figure 5.14 shows a picture of an IP-PBX whose functionality consists of connecting both to the IP network and to the PSTN network, in addition to having lines for internal telephone extensions.

Figure 5.15 shows connectivity of the IP-PBX to the IP network and to the PSTN network. This means that we can make calls between telephone devices that are directly connected to the IP network or make calls between IP network equipment and the PSTN network or among equipment in the PSTN network. As we can see once again, once the quality parameters for voice, video, and data transfer over IP have been correctly designed and implemented, it is necessary to appropriately configure the equipment to establish the service.

We will now see an example of configuration of an IP-PBX OmniPCX Alcatel. The purpose is merely to show what a configuration scheme looks like. Figure 5.16 shows the IP configuration of both the management card and the VoIP card.

Figure 5.17 shows configuration of local extensions administered by the IP-PBX.

Figure 5.18 shows configuration of DiffServ with the EF value for quality of service management.

Figure 5.19 shows IP configuration of the IP-PBX with which the corresponding telephone calls will be made in this example.

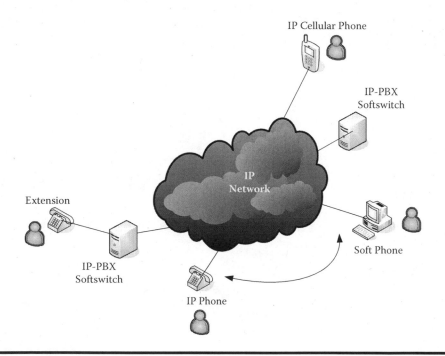

Figure 5.13 VoIP communication.

Now that we have explained integration of the IP-PBXs to the IP networks, we will discuss how cellular telephone has also become integrated with the IP network and how it is expected that cellular telephony devices will eventually feature IP addressing in order to make calls directly through this network.

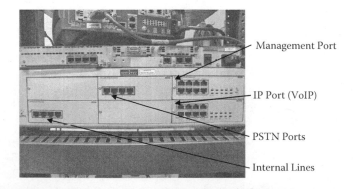

Figure 5.14 IP-PBX.

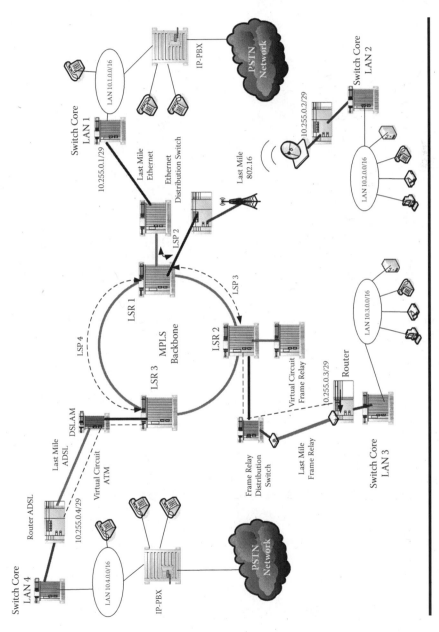

Figure 5.15   Network design for VoIP.

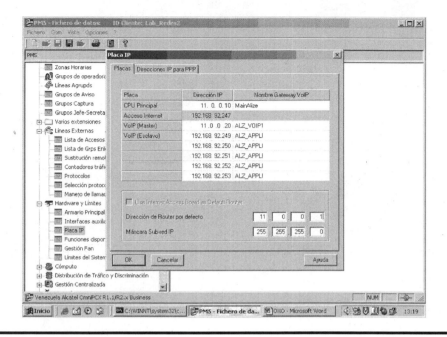

Figure 5.16 IP configuration.

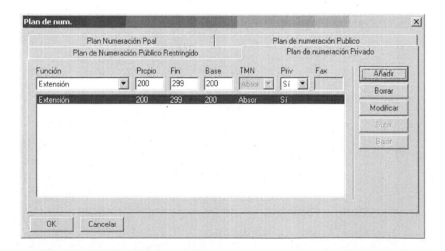

Figure 5.17 Local extensions.

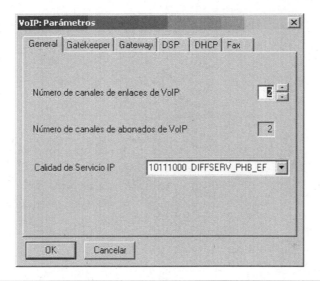

Figure 5.18   DiffServ configuration.

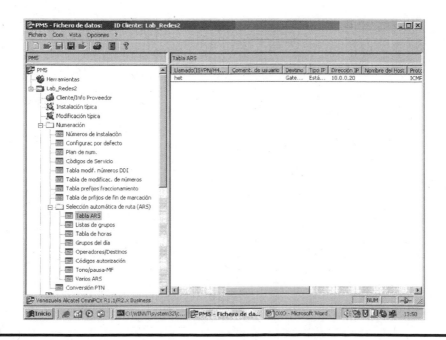

Figure 5.19   IP configuration of the remote IP-PBX.

5.2.1.2 Cellular IP

As cellular networks have evolved, we have witnessed that part of the concern is to increase capacity of data transmission over this network. On this matter we can say as an example that a platform called *General Packet Radio Service* (GPRS), also called 2.5G, was created with 2G networks. This platform allowed and increased transmission capacity over data networks and even the capability of transmitting data to and from the Internet.

Below is a brief summary of the evolution of cellular networks.

1G (or 1-G) is the abbreviation for the wireless technology for the first generation of cellular phones. The standards for analog cell phones were introduced in the 1980s and lasted until they were replaced by the second generation (2G) digital cell phones. The main difference between these two successful mobile telephony generations, 1G and 2G, is that the radio signals used by 1G are analog while the 2G networks are digital.

But, it is worth explaining here that both systems use digital signaling to connect the radio towers (which hear the mobile stations) to the rest of the system. But the call itself is coded in digital format in 2G while 1G is modulated only at a higher frequency (typically 150 MHz and higher). Systems considered *first generation* use modulated frequency (FM) for voice transmission and frequency shift keying (FSK) for signaling. The access option to the radio channel was Frequency Division Multiple Access.

First generation mobile telephony first appeared in 1979 and flourished during the 1980s. It introduced cellular phones based on cellular networks with multiple base stations, located relatively closely to each other, and protocols for "transfer" among cells when the telephone moved from one cell to another. Analog transfers and strictly for voice are the characteristics that identify this generation. With very reduced link quality, connectivity speed was not higher than 2400 bauds. Transfer among cells was imprecise and capacity was low due to the FDMA technique, which notably constrained the number of users that could simultaneously access the system, as protocols for stationary channels assignment suffer this constraint.

September 1981 saw the appearance of the first cellular telephony network with automatic roaming in Saudi Arabia, a system of the Nordic Mobile Telephone (NMT) Company. One month later, the Scandinavian countries began NMT networks with automatic roaming between countries. NMT is a standard used in countries such as Switzerland, the Netherlands, Eastern Europe, and Nordic Russia, while TACS was implemented in the United Kingdom; C-450 in the German Federal Republic, Portugal, and South Africa; RAdiocom-2000 in France; and RTMI in Italy. Japan had multiple systems. Three standards, TZ-801, TZ-802, and TZ-803, were developed by NTT, while a complete system component driven by DDI used the Japanese Total Access Communication System (JTACS) standard.

Of the technologies developed in the first generation of mobile telephones, it is worth discussing the Advanced Mobile Phone System (AMPS) more in depth, because it was the most transcendental. The architecture, components, and other parameters of the European networks are similar to those used by AMPS, only with some small differences in the frequency spectrum used and in the channels implemented.

5.2.1.2.1 Advanced Mobile Phone System (AMPS)

AMPS is a first-generation mobile telephony system developed by Bell Laboratories in the United States. It was first implemented in 1982.

5.2.1.2.1.1 General Operation — AMPS and same-type mobile telephony systems divide the geographical space into a network of cells or simple cells (hence, the name *cellular telephony*) in a way that adjacent cells never use the same frequencies to prevent interference. Each cell's base station emits with relatively little power, compared to the hundreds of watts of previous systems such as IMTS. The power emitted is less because the cells are smaller. A small cell size also favors reuse of frequencies and largely increases system capacity. However, a larger number of bases is also required and, consequently, a larger investment.

To establish communication between users occupying different cells, all base stations are interconnected to a mobile telephony switching office (MTSO), also called a *mobile switching center* (MSC). A hierarchy, like the one in regular telephony systems, is established. The use of cellular systems has certain problems, such as when users switch cells while talking. AMPSs foresee this and are able to always maintain active communication, provided there are available channels in the cell entered. This cell handoff is based on analyzing the power of the signal emitted by the mobile phone and received at the different base stations, and is coordinated by MTSO. Depending on the mode in which done, communication may be disconnected for approximately 300 milliseconds and restored immediately after or is imperceptible to users. These were subsequently called *hard-handoff* and *soft-handoff*, respectively.

5.2.1.2.1.2 Frequency Assignment — In 1980, the FCC allowed two common carriers per cellular service area. The idea was to eliminate the possibility of a monopoly and provide the advantages usually provided by a competitive environment. Next came two frequency assignment systems, A and B, each with its own group of channels that shared the assigned frequency spectrum. System A was defined for companies without telephone cables and system B for companies with cables.

The FCC originally assigned a 40 MHz frequency band to the AMPS system, formed by 666 duplex channels per service area, with a distance of 30 KHz between adjacent channels. Channels in system A range from 1 to 333 and system B channels range from 334 to 666. For mobile units, channel 1 has a transmission frequency of 825.03 MHz and channel 666 has 844.98 MHz. In cellular telephony systems, transmissions from base to mobile unit or from mobile unit to base station, simultaneously, are frequently necessary or preferable and are achieved with the full-duplex transmission mode. Duplex operation is achieved thanks to frequency of time domain methods. Frequency Division Duplication (FDD) provides two different band frequencies to every user. In FDD, every duplex channel actually consists of two simple channels (in one sense). A special device called a *duplexer* is used in every mobile unit and is also used in the base station to allow simultaneous transmission and reception of a duplex channel.

AMPS channel spectrums are divided into two sets or basic groups. One set of channels is dedicated to the exchange of control information between the mobile units and the cell site, and are called *control channels*. The second group, called *voice or user channels*, consists of the remaining channels and is used in real conversations and the exchange of data among users. Added frequencies are called *expanded spectrum* and include channels 667 to 799 and 991 to 1023.

Even though AMPS was replaced by digital systems such as GSM and D-AMPS, it was a system of vital historical importance for the development of mobile communications because of its success and for the novel ideas it contributed. Many operators currently still use it as a backup technology, as it covers more territory than the future digital technologies (TDMA, GSM, and CDMA). However, by being purely analog, AMPS is not compatible with short text messaging, SMS, or with any type of data.

5.2.1.2.2 Second Generation (2G)

Given the inefficiencies in the multiple access method, FDMA, used in first-generation (1G) cellular telephony, new alternatives were suggested to increase the number of simultaneous users in one cell. The second generation (2G) arrived in the 1990s, and, contrary to the former generation, was characterized by being digital. Digitalization brought with it a reduction in size, cost, and power consumption of the mobile devices, hence allowing cellular phone batteries to last longer. Other benefits of digitalization in cellular telephony were the services provided, such as caller identification, short message texting, SMS, voice messages, and multiparty conversations, just to name a few. These services depend on the service providers and the functions supported by the phone device. In addition, the new services extended their coverage to special environments such as highway tunnels, underground parking, inside buildings, etc. Characterization of the broadband channel also became increasingly important for operation optimization.

The four standards of second generation—GSM, TDMA, CDMA, and PDC—dominate wireless networks today, with hundreds of millions of users around the world, in contrast to the increasingly reduced number of users who are still connected to first-generation analog systems.

The 2G systems use more sophisticated coding protocols that are still used in present-day cellular telephony systems. The predominant technologies are Global System for Mobile Communications (GSM), standard IS-136 (TIA/EIA136 or ANSI-136, based on Time Division Multiple Access), CDMA (based on Code Division Multiple Access), and Personal Digital Communications (PDC), used in Japan. Protocols used in 2G systems support higher information speeds by voice, but are limited in data communication. Auxiliary services such as data, fax, and text messaging can be offered. Most 2G protocols provide different levels of encryption.

A new frequency band was opened to allow new services and new competitors. Band 1.9 GHz (1850–1990 MHz), known as *PCS* (Personal Communications Service), created new expectations for users and many telephone companies used the term *PCS* for their advertising campaigns.

PCSs are basically categorized in three technologies: GSM, TDMA IS-136, and CDMA IS-95. These three technologies are the predominant technologies in second generation.

GSM was the first commercial digital cellular system to be operated. It was developed in the 1980s from an initiative of the European Commission with equipment manufacturers for regional harmonizing of cellular networks. GSM uses TDMA access mode in the bands 900/1800/1900 MHz. This standard has also been adopted by all European countries; it is also very popular on other continents, with over 45% of the world's mobile subscribers (in April 1999).

TDMA IS-136, also known as *Digital AMPS System* (D-AMPS) is the evolution of AMPS used in the band from 800 to 1900 MHz. TDMA was introduced to protect the investments that service providers had made in AMPS technology, so that the transition from one generation to the next among its users could be accomplished using their existing networks, without bringing about inconveniences to subscribers. D-AMPS, like GSM, also expanded globally. TDMA IS-136 divides the 30 KHz bandwidth in three time slots, increasing capacity threefold with respect to the AMPS analog version. TDMA IS-136 allows speeds of up to 9.6 Kbps to incorporate its wide variety of digital services.

The TDMA Interim Standard 54 System (IS-54) was released by the Telecommunications Industry Association (TIA) in early 1991. In 1994, the FCC in the United States announced that it would assign spectrum for the PCSs in the 1900 MHz band and a series of auctions for that band began.

The personal digital cellular network (PDC) is the world's second digital standard for mobile systems, although it is used exclusively in Japan, where it was introduced in 1994. PDC is based on TDMA in the bands from 800 to 1500 MHz and operates at speeds of up to 9.6 Kbps. This system was introduced by Japanese operator NTT-DoCoMo as a substitute for the older analog systems.

CDMA is a specific form of technology known as spread spectrum and dates back to the 1940s, when it was used for military communications due to its immunity to interference and high security. In the late 1980s and early 1990s, the company Qualcomm developed a cellular system based on CDMA. In 1993 it was modified and adopted by TIA as Interim Standard 95 (IS-95), also known as *cdma-One*. Many operators adopted this standard in the bands from 800 to 1900 MHz. CDMA IS-95 (cdmaOne) supports data services on switched circuits at speeds of 9.6 to 14.4 Kbps. Protocol IS-95A supports speeds of up to 14.4 Kbps. Protocol IS-95B (based on packet switching), which was a transition step toward the next generation, provides data speeds of up to 64 Kbps, maintaining compatibility backwards with existing IS-95A systems. With CDMA, capacity of the system increases 10 to 15 times, compared to AMPS, and more than three times compared to the systems based on TDMA.

CDMA networks provide a high-speed wireless transmission of data capacity that provides clients information and image services no matter their location. This generic technology appears as the technological base of excellence for the next generation of mobile communications, 3G, being a 2G technology. In fact, the global trend in the industry is the adoption of CDMA.

Another digital system used in Japan is the Personal Handphone System (PHS). This standard, launched in 1995, is not a cellular technology per se. This technology was designed for densely populated areas where cellular systems may experience coverage problems, with lower costs than cellular telephony. By March 1999, there were approximately 5.77 million PHS users.

Implementation of second-generation systems began around 1991 when the first GSM networks became commercially available in Finland.

Some of the migrating pathways toward third generation suggested a gradual adoption of the technology. Thus appeared 2.5G systems, which exhibit better characteristics than 2G systems but do not strictly meet the requirements of 3G systems defined by the International Telecommunications Union (ITU).

Second-generation signaling adapts to protocol SS7 more efficiently and with greater capacity than do first-generation systems, and facilitates interconnection with the Integrated Services Digital Network (ISDN). In the case of GSM, there has been an evolution from a first basic network, which consisted of merely providing telephony service, to more complex networks, to which progressively have been added value services, provisions of intelligent network, and improved data transfer. Next, we discuss the structure of the GSM network.

5.2.1.2.3 Global System for Mobile Communications (GSM)

5.2.1.2.3.1 Novelties Introduced by the GSM System — The GSM system allows connection to the switched telephony network and the Integrated Services Digital Network, and allows offering telephony users data transmission (up to 9.600 bauds), group III facsimile, connection to electronic mail systems (X-400), and short message delivery (alphanumeric) that allows delivery and receipt from a mobile terminal, reading them in this latter case, in the corresponding visor. In addition, it supports other services such as call forwarding, incoming or outgoing call barring, three-party line, call waiting, and more. The security issue in this service provides important novelties, thanks to the use of the user card, which authenticates validity of the call; encryption facilitates with complete confidentiality (voice, data, and subscriber identity); and makes it impossible to use stolen equipment by prior assignment of a serial number to each mobile station.

Its radio component uses frequency bands from 900 MHz with the TDMA method, which provides eight telephone channels in one carrier and voice coding at 13 Kbps, assigning an eighth of time to every channel.

5.2.1.2.3.2 GSM Structure — In its basic structure, GSM is organized as a network of continuous radio electrical cells that provide complete coverage of the service area. Every cell belongs to a base station (BS or BTS) that operates a set of radio channels different from those used in adjacent cells and which are distributed according to a cellular plan. One group of BSs is connected to a base station's controller (BSC), which is responsible for things such as handover (transfer of the mobile from one cell to another) or control of BS power and the mobiles. Consequently, BSC is responsible for managing the complete radio networks and represents a true novelty with respect to the previous cellular systems. One or several BSCs connect to a mobile switching center (MSC). This is the heart of GSM, responsible for initiating, routing, controlling, and finalizing calls, as well as for the tariff information. It is also the interface among varied GSM networks or between one of them and the public telephony or data networks.

Information regarding subscribers is stored in two databases known as the *home location register* (HLR) and the *visitors location register* (VLR). The first database analyzes the levels of subscription, supplementary services, and current or most recent location of mobiles in the local network. The authentication center (AUC) works in association with HLR; it contains the information by means of which authenticity of calls is verified to avoid possible fraud, the use of stolen SIM cards, or enjoying unpaid service. The VLR contains information regarding subscription level, supplementary services, and location area of subscribers who are or at least were in another visited zone. This database also contains information regarding whether subscribers are active or not, thus preventing unproductive use of the network (sending signals to a disconnected location).

5.2.1.2.3.3 Equipment Identification Register — The equipment identity register (EIR) stores information about the type of mobile station being used and can prevent calls when it identifies that the call has been stolen, belongs to a nonrecognized model, or experiences failure that might negatively affect the network. As for network communications, a new digital signaling scheme has been developed. For communication between MSCs and location registers, one uses the mobile application part of signaling system number 7 of the CCITT, an indispensable formula for international operation of GSM networks.

5.2.1.2.3.4 The GSM Radio Interface — GSM uses two multiplexing techniques: FDMA and TDMA. FDMA consists of dividing the assigned bandwidth to the cell into radio channels. One radio channel is a pair of frequencies: one for the uplink between the mobile and the BTS, called *Fu*, and the other for the downlink between the mobile and the BTS, called *Fd*.

Total bandwidth for the GSM system is from 890 to 915 MHz (25 MHz) for the uplink and from 935 to 960 MHz (25 MHz) for the downlink. Every radio channel fills a bandwidth of 200 KHz in every sense and, therefore, there will be a total 125 uplink carriers and 125 downlink carriers; one pair is used for control. Every cell shall have a subset of such carriers. A subdivision of smaller channels takes place within every radio channel using TDMA. In other words, time is divided into time intervals or slots, so that they can have eight subchannels that will form a segment with a duration of 4.62 milliseconds (ms).

Some of these subchannels are used to transfer user information and others for signaling and control tasks. Transmission of the information is done by bursts, every 4.62 ms. Hence, for example, the mobile transmits a burst of information every 4.62 ms that must be introduced into an assigned time slot for communication in the uplink. Similarly, it will receive information every 4.62 ms in the time slot it has been assigned in the downlink.

5.2.1.2.4 CDMA Technology

CDMA is a generic term that defines a wireless air interface based on spread spectrum technology. For cellular technology, CDMA is a multiple access technique specified by the TIA as IS-95. In March 1992, TIA established subcommittee TR 4.5 with the purpose of developing a digital cellular telephony standard with spread spectrum. In July 1993, TIA approved standard CDMA IS-95.

One of the unique traits of CDMA is that despite the existence of a limited number of phone calls that can be manipulated by a carrier, it is not a fixed number. Capacity of the systems depends on many factors. Every device that uses CDMA is programmed with a pseudocode, which is used to expand a low-power signal over a broad frequency spectrum. The base station uses the same code but inversely to filter and reconstruct the original signal. The other codes remain expanded, distinguishable from the background noise.

There are many variations today, but the original CDMA is known as *cdmaOne* under a Qualcomm registered trademark. The CDMA is characterized by its high capacity and small radius cells, which use spread spectrum and a special coding scheme, and for its efficient use of power.

5.2.1.2.4.1 Standard CDMA (IS-95) — Standard IS-95 has been defined by the U.S. TIA and is compatible with the existing frequencies plan in the United States for analog cellular telephony. The bands specified are 824–849 MHz for reverse-link and 869–894 MHz for forward-link. The channels are separated by 45 MHz. Maximum user speed is 9.6 Kbps and a 1.2288 Mbps channel expands. The expansion process is different for every link. In the forward-link data are coded with a convolution code (1/2), interleaved, and expanded at a sequence of 64 bits (Walsh functions). Every mobile is assigned a different sequence. In addition, a pilot channel (code) is provided for each mobile to determine how to act with respect to the base. This channel has higher power than the rest and provides a coherent base that is used by mobiles to demodulate traffic. It also provides time reference for code correlation. A different scheme is used in the reverse-link because the data can reach the base by very different routes. Data are coded with a convolution code (1/3). After being interleaved, every six-bit block is used as an index to identify a Walsh code. Finally, the signal is expanded using the user-specific and base-specific codes. The power control takes place in the steps from 1 dB and can be done in two ways: one is to take as reference the power received from the base station; the other is to receive instructions of base on the adjustment that must take place. Lastly, it must be noted that the signal being transmitted is modulated using the QPSK technique filtered from the base to the mobile and QPSK filtered with a displacement from mobile to base.

5.2.1.2.5 D-AMPS

Digital–Advanced Mobile Telephony Service (D-AMPS) was the product of the technological evolution of its predecessor, the system AMPS, so that first-generation and second-generation mobile telephones could operate simultaneously using the same physical infrastructure as AMPS cells. It uses the same 30 KHz channels as AMPS and the same frequencies. Depending on the combination of telephones in the cells, the MTSO of the cell determines which channels are analog and which are digital, and can change the types of channels dynamically, as the combination of channels changes in a cell.

Uplink channels are in the 1.850 to 1.919 MHz range and downlink channels are in the 1.930 to 1.990 MHz range, again, in pairs.

5.2.1.2.6 2.5G Generation

There is no technological standard that can be called *2.5G* or *2.75G*, but certain mobile networks called *2G* incorporate some of the improvements and technologies of the 3G standard in 2G networks and with data transfer rates higher than those of regular 2G systems but lower than 3G.

Once second generation was established, the constraints of certain systems regarding delivery of information became evident. Many applications for information transfer were seen as the use of laptops and the Internet became increasingly popular. Whereas third-generation was on the horizon, some services became necessary before its arrival. General Packet Radio Service (GPRS), developed for the GSM system, was one of the first to be seen. Until that time, all circuits were dedicated exclusively to every user in the switched circuit scheme. This was inefficient when a channel transferred only a small percentage of information. The system allowed users to share the same channel, directing information packets from the source to the destination, to allow more efficient use of the communications channels.

Subsequently, improvements were made to the information transfer rates with the introduction of the Enhanced Data Rates for Global Evolution (EDGE) system. This system is basically the GPRS system with a new frequency modulation frequency scheme. While GPRS and EDGE were applied to GSM, other improvements were aimed at CDMA, the first step being from CDMA to CDMA2000 1x.

2.5G provides some of the benefits of 3G (for example, data switching in packets) and can use some of the infrastructure used by 2G in the GSM and CDMA networks. Whereas the terms *2G* and *3G* are officially defined, *2.5G* is not. It was invented for advertising purposes only. Many of telecommunications service providers will move to the 2.5G networks before massively entering 3G. 2.5G technology has been the fastest and most inexpensive way to update to 3G.

5.2.1.2.6.1 GPRS Technology — Existing data services in 2G cellular networks were not meeting user and provider needs. From the user point of view, data rates were too slow and connectivity configuration was too long and complicated. The service was also too expensive for most users. From a technical point of view, the disadvantage was the fact that 2G wireless data services are based on switched circuits in the radio transmission. In the air interface, a full traffic channel is assigned to a single user for a complete call period. In the case of burst traffic (i.e., Internet traffic), this results in a highly ineffective use of resources. It is evident that for burst traffic, services that support switched packets provide better use of traffic channels. This is true because a single channel will be assigned only when necessary and will be released immediately after the packet transmission. With this principle, multiple users may share a single physical channel.

In order to correct such deficiencies, two cellular data packet technologies have been developed: Cellular Digital Packet Data (CDPD), for AMPS, IS-95, and IS-136, and General Packet Radio Services (GPRS). GPRS was originally developed for GSM, but will also be integrated with IS-136. The next section discusses GPRS from a GSM point of view.

GPRS is a new GSM-supportable service that largely improves and simplifies wireless access to data packet networks, i.e., Internet. It applies the radio packet principle of transferring user data efficiently between the GSM mobile stations and external data packet networks. Packets may be routed directly from the GPRS mobile stations to the switched packet networks. Networks based on the Internet Protocol (IP), for example, global Internet, or on corporate intranet networks are supported on the current version of GPRS. GPRS users benefit from shorter access times and higher transfer rates. In conventional GSM, configuration of the connection takes several seconds and transmission rates are restricted to 9.6 Kbps. In practice, GPRS provides session establishing times under one second and ISDN-type transfer rates of several tens of Kbps.

5.2.1.2.6.2 EDGE — Enhanced Data Rates for Global Evolution (EDGE), is an update of GPRS that uses the most of the investments made in packet infrastructure and circuit switching by reutilization. Most of the changes in EDGE deployment focus on radio infrastructure, where software and hardware are added in every cell.

5.2.1.2.7 Third Generation (3G)

Progress made in third-generation systems by the ITU in the late 1980s were in the beginning called the *Future Public Land Mobile Telecommunications System* (FPLMTS). The name was finally changed to *International Mobile Telecommunications System* (IMT-2000), and was created with the purpose of assessing and specifying the requirements of 3G cellular standards to provide high-speed data and multimedia services.

5.2.1.2.7.1 Requirements of a 3G System — The following requirements are contained in ITU and other standardization bodies' recommendations, without which a system should not be considered third generation:

- High-speed data transmission of up to 144 Kbps for data from mobile users (vehicle), up to 384 Kbps for portable data (pedestrian), and up to 2 Mbps for fixed users (stationary terminal)
- Symmetric and asymmetric transmission
- Packet switching services, such as Internet traffic (IP) and real-time video, with connectivity with switched circuits
- Voice quality comparable to the quality provided by wireless systems
- Greater capacity and better efficiency of the spectrum compared with the previous systems (1G and 2G)
- Compatibility with second-generation systems and possibility of coexisting and interconnecting with satellite mobile services
- International roaming between different operators

Third-generation systems must provide support for applications such as voice in narrow band to real-time multimedia services and bandwidth; capacity to manage data at high speed for network surfing; delivery of information such as news, traffic, and finances by means of pushing techniques; and remote wireless access to Internet and intranets, unified message services such as e-mail, applications of mobile e-commerce, which include banking transactions and mobile purchases, real-time audio/video applications such as video telephone, active videoconference, audio, and music, specialized multimedia applications such as telemedicine, and remote security supervision.

5.2.1.2.8 Institutions Involved in the Development of 3G Systems

In Europe, the European Telecommunications Standards Institute (ETSI) proposed the Universal Mobile Telecommunications System (UMTS) third-generation paneuropean standard. UMTS is a member of the global IMT-2000 family of the third-generation mobile communications system of the UIT. In the United States, the American National Standards Institute (ANSI) worked on the evolution of the AMPS/IS-136 and CDMA/IS-95 systems. In Japan the Association of Radio Industries and Businesses (ARIB) also worked on CDMA to prepare third-generation standards.

The regional standardization entities ETSI (Europe), ANSI (United States), ARIB (Japan), and TTA (Korea) worked on separate proposals for the W-CDMA standard. These regional entities joined efforts in the 3G Partnership Project (3GPP), and today there is a joint standard, W-CDMA. ITU received three proposal families for FDD (WCDMA, CDMA 2000, and UWC-136) and three proposals for TDD (UTRA/TDD, TDD-SCDMA, and DECT). Efforts have subsequently been coordinated to harmonize IMT-2000 candidates and finally have availability to the 3G compressed standards.

5.2.1.2.9 Fourth Generation (4G)

The main proposal of fourth generation (4G) is the implementation of Voice over IP protocol from the cellular phones themselves. The communications protocol that will enable this new convergence is Unlicensed Mobile Access (UMA). This technology allows users to connect wirelessly to the Internet and to make calls through it.

At present, one can make calls over the Internet only using a computer; the device turns into the link of calls over the Internet. With UMA it will be possible to do the same with a mobile phone instead of a computer. This will enable users to connect to a broadband wireless network and make their calls over the Internet. Not only will they be able to talk on the phone, but they will be able to access all of the GSM/GPRS services of their cellular network: surf the Web, access e-mail, MMS, SMS, data download, and other possibilities. With UMA technology, an operator may provide connection in places without coverage, as it may be able to use a Wireless Local Area Network (WLAN) to allow its users to connect to the Internet and make calls; the purpose is to get third-generation interaction with short-range technologies, as is the case in a WLAN. The cellular phone that will operate both in the cellular network and the wireless WLAN will be, according to experts, a four-band telephone. This means that it will work with bands of the cellular network and of the WLAN infrastructure.

This technology—already available in several parts of the world—will drive current users to replace fixed telephones with cellular telephones and leave fixed telephony. When 4G is officially implemented will depend on regulation management.

5.2.1.2.9.1 Introduction of IP in Cellular Mobile Networks — With the help of Internet, IP protocol is becoming the common language of current communications networks. However, adopting this protocol for certain applications is not free of problems, many of which are still pending resolution.

Evolution of current second-generation (2G) networks toward third-generation (3G) seeks to provide increasingly elevated speeds and access via switching packets, as is the case with GPRS, in order to prepare for a horizon where data traffic shall surpass voice traffic, with access to IP networks in general and to Internet, specifically, being the main cause of this situation. If one adds to all this that the work being performed to support Voice over IP carrying is fairly mature, with available commercial solutions, one can understand why the option of a cellular communications access network fully based on IP has gained strength.

The next logical step would be for this network to naturally assume all the functions necessary to support mobility. Two groups that develop 3G are moving in this direction: the Third-Generation Partnership Project (3GPP) and the Third-Generation Partnership Project 2 (3GPP2). They have each individually studied how to convert their access networks into IP networks totally based on routers and how to adopt the work being performed by the Internet Engineering Task Force (IETF) to provide mobility via IP. The IETF has been working on a universal solution to achieve IP mobility known as *Mobile IP*. This is a general solution that has not been optimized for any type of access network and, therefore, has serious inconveniences. One of the key requirements that 3G networks, and overall cellular networks, demand is support of micromobility (the possibility of changing access points frequently and rapidly within the network). Mobile IP has serious constraints to comply with this requirement.

There currently exists important synergy between 3G groups and IETF: on one hand, one is attempting to improve the Mobile IP protocol to comply with 3G requirements, without losing sight of its universal solution character, and, on the other hand, one has created a new micromobility specialty group. Leaving IETF and the 3G groups aside, it is also important to mention the Mobile Wireless Internet Forum (MWIF). Created in February 2000, this forum consists of companies of great significance in mobile communications, data networks, software, and electronics. Its purpose is to develop key specifications that enable the use of IP in any type of wireless network, seeking to group in a single approach the efforts of other groups such as IETF, 3GPP, and 3GPP2. A general description of two of the most important technologies based on IP—Mobile IP and Cellular IP—is provided below.

5.2.1.2.9.2 Mobile Protocol IP — Mobile IP defines two new entities generically known as agents: the home agent (HA) and the foreign agent (FA), which are nothing more than two routers, one in the source network of the mobile node and another in the one it visits, and which perform data management functions similar to those of the HLR and VLR of the GSM cellular network. When a mobile node moves to another network, it receives a signal to register from the FA of the visited network. This way, it detects the change of location, whether with respect to the HA or to a previous FA. It then acquires the address of the FA, which remains recorded in its HA. From this point, any datagram sent to the mobile node passes through its HA, which forwards it to the FA via a tunnel that is responsible for getting it to the mobile node. Conversely, the mobile node sends the datagrams directly to their destination node.

The simplicity of the Mobile IP protocol has a price: supporting micromobility. In high-mobility environments such as cell phones, where the mobile node is frequently changing access points, performance of the protocol may not be as appropriate for the type of service one wants to support. Every change, whether within the same network, requires a signal exchange with the HA, which slows down the update process with the subsequent loss of packets this implies.

5.2.1.2.9.3 Cellular IP Protocol — Cellular IP is one of the protocols exclusively designed to support micromobility, and combined with Mobile IP it allows complete mobility within an IP network.

By being a cellular system, Cellular IP provides a series of advantages that, when applied correctly, may improve the provisions of future IP networks without losing any of the properties that characterize IP networks, such as flexibility, scalability, and robustness. It inherits from cellular systems the principles of mobility management, transfer control, and location of inactive nodes. Cellular IP is based on simple and inexpensive nodes that may be interconnected to form arbitrary topologies and operate without a previous complicated configuration.

The universal component of a Cellular IP network is the base station, which serves as radio access point to the network while concurrently forwarding IP packets and integrating control functions of cellular systems traditionally implemented in MSCs and BSCs. The IP forwarding is substituted by the Cellular IP's own forwarding, where location and support are integrated to the transfer without altering the IP protocol stack. Moreover, a base station may be configured to perform gateway functions. A gateway is the node responsible for connecting the Cellular IP access network to the Internet. When a mobile node originating in an external network is put into service, the gateway acts as a mobility agent (FA) of the Mobile IP protocol.

Base stations periodically emit signals called beacon signals. These signals incorporate information about the base station and the network. The mobile nodes use them to detect which is the closest base. A mobile node transmits it packets through the base station from which it receives the best quality signal. For the mobile node, this base station acts as its router by default. All IP packets sent by the mobile node are forwarded directly from the base station to the gateway, regardless of the destination address. To do this, a forwarding algorithm of the type *hop-by-hop shortest path* is used; that is, the packet follows the shortest path to the gateway.

All CIP nodes have a table called *route cache*. Packets sent by a mobile node update the data of these tables in all the nodes they pass. An entry in this cache saves the IP address of the mobile node that generated the packet, the network interface through which it passed, and the physical address of the last downlink neighbor through which it passed. The linking of all the information on these tables, referred to as a *mobile node*, serves to reconstruct the path required for packets addressed to such node to reach their destination. Even if a mobile node changes its access point to the network, the route cache tables point to the new node location, as mobile node is responsible for generating new entries before sending route update packets. These packets are also used to prevent mobile node entries from expiring, as they have a limited life period within the route caches. Thus, if a node wants to stay active and traceable on the network, but does not have packets to send that maintain its cache entries, it must periodically submit route-update packets.

Cellular IP also incorporates a paging mechanism that allows an inactive mobile node to maintain is traceability within the network. Other tables known as *paging caches* are used for this purpose. By consulting these tables, a node can forward packets toward inactive nodes whose information is not available in the routing caches.

Finally for this section, Figure 5.20 illustrates integration of a cellular network, whether third-generation EDGE (EGPRS) or UMTS, with the IP network. It also shows IP Cellular equipment. In this case, one could make telephone calls among equipment in the same network, whether cellular or IP or between an equipment in the UMTS network with another that is located in the IP network.

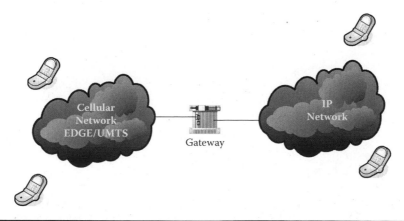

Figure 5.20 Cellular network with IP.

5.2.2 IPTV

Despite the fact that device manufacturing technologies have progressed, reducing prices, the use of cellular phones with technology for network, multimedia, etc., functions is not yet common. But there is a mass-use device in all homes: the television set. Until today, television has been a unidirectional medium, where one can only perform basic functions such as switching channels and increasing or lowering the volume. Data networks have also made great progress, providing an increasingly faster service. Today, not only is data transmitted over the Internet, but so are voice and video. This has resulted in the emergence of Television over Internet Protocol (IPTV).

IPTV does not imply going to Web sites to watch television programming. Transmission of TV over IP implies the method by which the information is sent, in other words, how video and voice are sent as IP packets until they reach the users.

IP television allows personalized programming for every user; in other words, they will receive only the programs they want, at the time and with the commercials they wish, contrary to conventional television, where programming is received by diffusion; that is, all users receive the same channels and programs.

Present-day television sets receive analog signals. For this reason, they require a device for IP conversion into analog video. This is the set-top box, which also performs an interactive function of the television over IP. TV signals are coded and converted into IP packets that will then be forwarded to the Internet. Next, they are distributed through the network to the final user who with a set-top box acquires the analog signals to submit to the television set. One important step in transmission of any type of digital television is video compression. Among the most common formats used are H.261, H.263, MPEG-2, and MPEG-4, the latter two being the most commonly used in IPTV.

IPTV is currently one of the most demanding IP services in terms of QoS. A good-quality IPTV transmission may demand between approximately 2 Mbps and 4 Mbps depending on the codec used, and may even demand higher transmission rates.

Figure 5.21 shows the connectivity scheme of IPTV service in an IP network. For this case, connecting a traditional TV set, one must use a device called a *setup box*. The setup box is a device that features an IP port and a coaxial port to connect it to the TV and hence project the signal in the TV set. Figure 5.21 shows the general scheme of an IP service transmission network. In this case, one can receive the original signal from a satellite television network (it could also be any other type of television network), a content server in which we could have pay-per-view or on-demand video, or a digital camera transmitting news or something similar. These signals are subsequently converted into IP packets and are traditionally transmitted via multicast addressing, although the service is also adapted for unicast. One also has the client who is receiving the IPTV service at home or the office. In this case the setup box will be the device responsible for receiving the IP packets to then send the signal

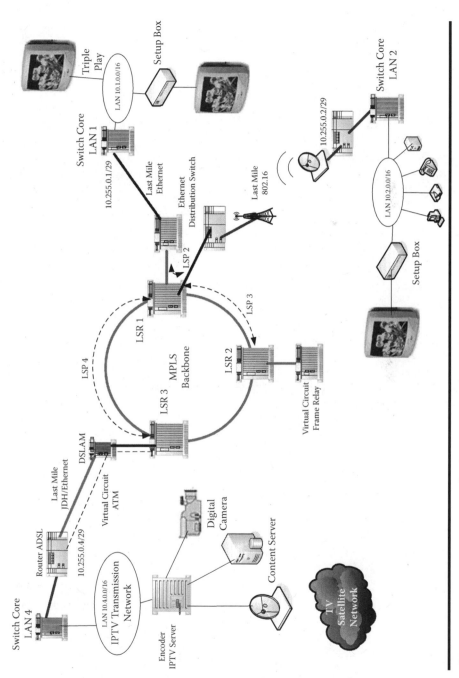

Figure 5.21 IPTV over IP network.

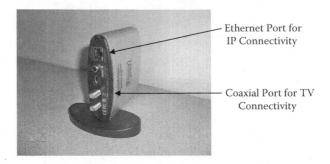

Ethernet Port for
IP Connectivity

Coaxial Port for TV
Connectivity

Figure 5.22   Setup box: Ethernet port for IP connectivity; coaxial port for TV connectivity.

to the TV set via coaxial cable or voice and video cables. Another option would be to receive the IP packets directly in the equipment, called *triple play*.

Figure 5.22 shows a setup box. The Ethernet port, with which it will be connected to the IP network, and the coaxial ports for connection to the TV set can be seen in this device.

5.2.3 Videoconference

Just as in the services previously mentioned, once the IP infrastructure and the QoS requirements for the different services have been implemented, any other new service may be set up in the network.

Another service that can be set up in the IP network is videoconference. This can be done just as for IPTV, as point-to-point communication or as multiple point–to–multiple point communication. This type of application is real-time and also requires certain QoS parameters, but videoconference is not as demanding on bandwith level as IPTV. A good videoconference may be done from 256 Kbps.

Figure 5.23 shows the connectivity scheme of the videoconference service in an IP network. As can be seen, what is needed for this operation is to know the IP address of each of the videoconference stations so that the calls to each respective IP address can be made to transmit voice and video. IP multicast addresses can also be used for this service.

Figure 5.24 shows a videoconference station. In this equipment one can see the Ethernet port to connect it to the IP network and the voice and video ports for connection to a video beam or a TV set.

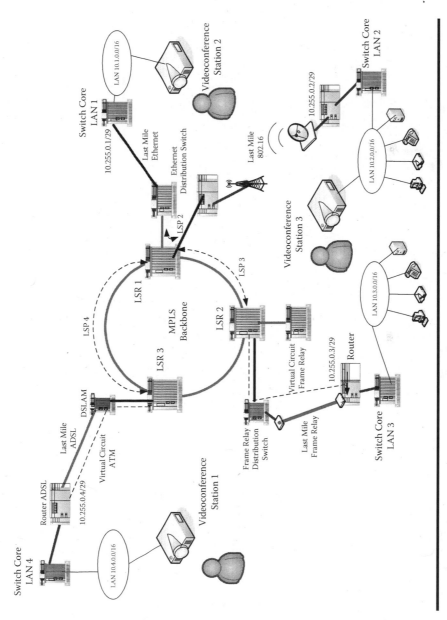

Figure 5.23 Videoconference over IP network.

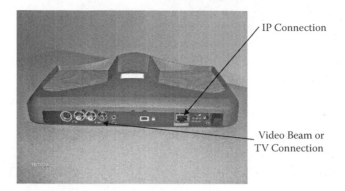

IP Connection

Video Beam or
TV Connection

Figure 5.24 Videoconference station: IP connection; video beam or TV connection.

5.2.4 *Video Streaming*

The final service to be discussed in this book is video streaming. There are other important services, such as the ones already mentioned, but they are not the intended purpose of this book.

The difference between a videoconference and video streaming is that the videoconference takes place online whereas video streaming transmissions take place offline or are at least processed through a server. In this case, video streams do not have as many QoS parameter requirements compared to other services such as IPTV, VoIP, and videoconference. In this case, the receivers or clients use a buffer to store the received IP packets and subsequently project them to the user, conveying the idea of online transmission.

Figure 5.25 shows the connectivity scheme of the video-streaming service in an IP network. As can be seen, for this service to operate, one needs to know the IP address of every receiver. It is also possible to use multicast IP addresses for this service.

With this last service discussed in this book, we show that many services may be integrated by IP convergence. The important thing is to define the appropriate IP structure and layer 1 and layer 2 platforms in the carrier backbones as in the last miles, and to establish adequate QoS parameters for each type of service. Once all of the foregoing has been done appropriately, one can simultaneously transmit different services, satisfying the needs and expectations of clients.

5.3 Introduction to NGN and IMS Networks

Having designed the network infrastructure to support convergence of services with quality, in this section we will present some fundamentals of Next Generation Networks (NGNs) and IP Multimedia Subsystems (IMSs). This section will

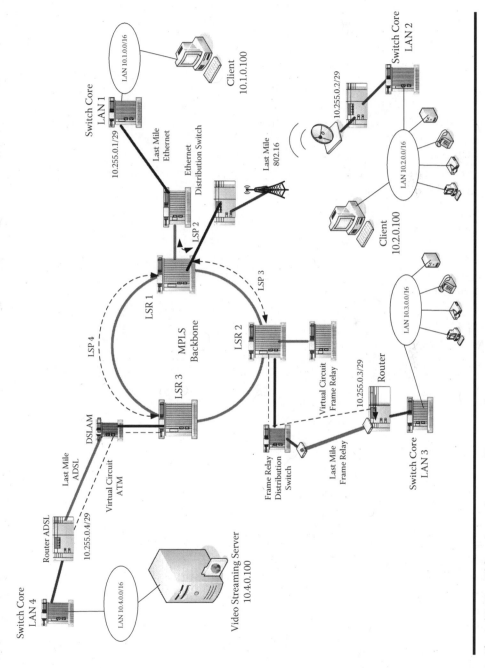

Figure 5.25   Video streaming over IP network.

summarize the behavior and objectives of these new networks, but will not discuss their architecture as there are other good books we recommend to understand the design and architecture of these networks, which are, obviously, based on everything we have discussed in this book,

NGN refers to the evolution of the current telecommunications and telephone access networks infrastructure, with the purpose of achieving convergence of the new multimedia services (voice, data, video, etc.) over the same network infrastructure. The main idea behind these types of networks is the carrying of encapsulated information packets through the Internet. These new networks are constructed from the Internet Protocol (IP); the term *all-IP* is commonly used to describe this evolution.

According to the ITU-T, a Next Generation Network is a network based on the transmission of packets with the capability of providing integrated services, including traditional telephone services, and of exploiting to the maximum the channel's bandwidth using the quality of service technologies so that transportation is completely independent from the infrastructure of the network used. In addition, it provides free access to users of different telephone companies and supports mobility, allowing multipoint access to users. From a more practical point of view, Next Generation Networks imply two essential changes in the traditional network architecture that have to be evaluated independently.

Regarding network backbone, NGN implies consolidation of several carrying networks (dedicated or overlay) constructed from different individual services (normally based on IP and Ethernet protocols). It also implies, among many other things, migration of the voice service from the traditional switched architecture (PSTN) to the new VoIP in addition to the substitution on the traditional networks (legacy-service) such as X.25 or Frame Relay. This even implies a migration of the traditional user to a new service such as IP VPN or transformation of traditional networks' infrastructure.

Regarding access networks, NGN implies migrating from the traditional dual voice and data channel associated with xDSL networks to convergent installations in which DSLAMS integrate voice or VoIP ports, thus leaving behind the current switched networks that multiplex voice and data through different channels.

Separation is well-defined in the NGNs between what corresponds to the carrying networks (connectivity) and the services running over such network. This means that every time that a telephone provider wants to enable a new service, he can easily do it by defining it from the service layer directly without taking into account the carrying layer. As has already been said, prorated services shall be independent of the network infrastructure. The current trend is for these services, including voice, to lean toward network independence and normally reside in the users' devices.

NGNs are based on Internet technologies, including IP protocol and MPLS. In the application layer, SIP protocols seem to have incorporated from standard ITU-T H.323. Initially, H.323 was the most popular protocol, although its popularity has decreased with time. For this reason, while traditional VoIP services are being developed, the new SIP services are being received better. However, while in voice networks full control is under the telephone operator, most of the large-scale carriers use H.323 as the best choice. Nevertheless, most carriers are conducting an intensive survey and lend support toward IMS, which gives the SIP protocol a better opportunity for becoming the new most widely used protocol. One of the most important devices in NGN for voice applications is softswitch, a programmable device that controls Voice over IP (VoIP) calls. This device enables correct application of the different protocols in the NGN. Its most important function is to create the interface for the current PSTN telephone network through the signaling gateways and the media gateway. However, *softswitch* may be defined differently by component manufacturers and may even have more and different functionalities. IP Multimedia Subsystem (IMS) is a standardization of NGN architecture for Internet multimedia services defined by ETSI and 3GPP.

An IMS network is an architecture per layers developed primarily by 3GPP to allow provision of the multimedia services currently provided by the Internet, both to mobile as well as fixed users that are part of a convergent IP network. In other words, it has multiple access types and provides multiple services, all over Internet Protocol (IP).

It is considered a multimedia subsystem because it provides real-time multimedia services (video, voice, videoconference, online games) and not real-time services (data, messages [SMS, MMS], mobile TV, Web, images, shared boards, push-to-talk)—in general, a set of tools that allow users to use any of these anytime/anywhere features, as is the case with mobile cellular telephony—but with a 3G hand-held device, using the carrier's IP backbone, and using packet switching, which is more efficient than circuit switching.

From an infrastructure point of view, IMS defines a clear separation between the carrying layers, session control, and application. Thus far, vertical architectures force the use of many exclusive elements and capabilities that complicate the creation and deployment of new services. Instead, IMS adopts the concept of layer architecture and expands it within a horizontal architecture in which the services and capabilities may be reused by different applications. The layers are from the bottom up: access, transportation, control, service, or application.

The access layer may represent all high-speed access such as UMTS Terrestrial Radio Access Network (UTRAN), CDMA2000 (broadband access technology used in the mobile networks in the United States), xDSL, cable networks, Wireless IP, WiFi, etc. The transportation layer represents an IP network. This IP network may integrate service quality mechanisms with MPLS, DiffServ, RSVP, etc. The transportation layer consists of the routers (edge routers for access and core routers for transit) connected by a transmission network. Different transmission stacks may be contemplated for the IP network: IP/ATM/SDH, IP/Ethernet, IP/SDH, etc. The control layer consists of session controllers responsible for routing signaling among users and for invoking new services. Nodes called *Call State Control Functions* (CSCFs) are responsible for this. IMS, therefore, introduces session control over the packet field. The application or service layer introduces the applications (added value services) proposed to users. Operators may position themselves thanks to the control layer, as integrator of the services they provide or that are provided by third parties. The application layer consists of application servers (ASs) and Multimedia Resource Functions (MRFs), which provides all IP Media Substations (IMSs).

Finally, NGN and IMS networks require the network infrastructure and designs that we have discussed in this book and the quality of service configurations in order for convergence of such services to take place appropriately for each of the applications' requirements. It is with the adaptation of protocols SIP or H.323 operating over these network infrastructures that one can implement these NGN and IMS networks.

References

[ALW04] Alwayn, V., *Optical Network Design and Implementation*, Cisco Press, 2004.

[AND07] Andrews, J., Ghosh, A., and Muhamed, R., *Fundamentals of WiMax: Understanding Broadband Wireless Networking*, Prentice Hall, 2007.

[ASH02a] Ash, J., Girish, M., Gray, E., Jamoussi, B., and Wright, G., Applicability statement for CR-LDP, *RFC 3213*, January, 2002.

[ASH02b] Ash, J., Lee, Y., Ashwood-Smith, P., Jamoussi, B., Fedyk, D., Skalecki, D., and Li, L., LSP modification using CR-LDP, *RFC 3214*, January, 2002.

[AWD01] Awduche, D., Berger, L., Gan, D., Li, T., Srinivasan, V., and Swallow, G. RSVP-TE: Extensions to RSVP for LSP tunnels, *RFC 3209*, December 2001.

[BUC00] Buckwalter, J., *Frame Relay: Technology and Practice*, Addison-Wesley, 1999.

[CAM06] Camarillo, G., and Garcia-Martin, M-A., *The 3G IP Multimedia Subsystem (IMS): Merging the Internet and the Cellular Worlds*, 2nd Edition, Wiley, 2006.

[COM05] Comer, D., *Internetworking with TCP/IP*, Vol. 1, 5th Edition, Prentice Hall, 2005.

[DAV00] Davie, B., and Rekhter, Y., *MPLS Technology and Applications*, Morgan Kaufmann Publishers, 2001.

[DAV08] Davie, B., and Farrel, A., *MPLS: Next Steps*, Vol. 1, The Morgan Kaufmann Series in Networking, 2008.

[DE06] De Ghein, L., *MPLS Fundamentals*, CRC Press, 2006.

[DUA02] Duato, J., Yalamanchili, S., and Ni, L., *Interconnection Networks*, The Morgan Kaufmann Series in Computer Architecture and Design, 2002.

[ELB06] El-Bawab, T., *Optical Switching*, Springer, 2006.

[ELL03] Ellis, J., Pursell, C., and Rahman, J., *Voice, Video and Data Network Convergence: Architecture and Design, From VoIP to Wireless*, Elsevier, 2003.

[EVA04] Evans, S., *Telecommunications Network Modelling, Planning and Design*, IEE The Institution of Electrical Engineers, 2004.

[EVA07] Evans, J. W., and Filsfils, C., *Deploying IP and MPLS QoS for Multiservice Networks: Theory & Practice*, The Morgan Kaufmann Series in Networking, 2007.

[GHA03] Ghanbari, M., *Standard Codecs: Image Compression to Advanced Video Coding*, The Institution of Electrical Engineers (IEE) Telecommunication Series, 2003.

[GOL07] Goleniewski, L., and Jarrett, K., *Telecommunications Essentials: The Complete Global Source*, 2nd Edition, Addison-Wesley, 2007.

[GUI05] Guichard, J., Le Faucheur, F., and Vasseus, J-P., *Definitive MPLS Network Designs*, Cisco Press, 2005.

[HAR07] Harte, L., *IPTV Basics, Technology, Operation and Services*, Althos Publishing, 2007.

[HEL03] Held, G., *Ethernet Networks: Design, Implementation, Operation, Management*, Wiley, 2003.

[HEL06] Held, G., *Understanding IPTV*, Auerbach Publications, 2006.

[HEL08] Held, G., *Carrier Ethernet: Providing the Need for Speed*, CRC Press, 2008.

[IBE97] Ibe, O., *Essentials of ATM Networks and Services*, Addison-Wesley, 1997.

[IBE01] Ibe, O., *Converged Network Architectures: Delivering Voice and Data Over IP, ATM, and Frame Relay*, Wiley, 2001.

[JAM02] Jamoussi, B., Andersson, L., Callon, R., Dantu, R., Wu, L., Doolan, P., Worster, T., Feldman, N., Fredette, A., Girish, M., Gray, E., Heinanen, J., Kilty, T., and Malis, A., Constraint-based LSP setup using LDP, *RFC 3212*, January, 2002.

[KRA02] Krauss, O., *DWDM and Optical Networks, An Introduction to Terabit Technology*, Siemens, 2002.

[KUR07] Kurose, J., and Ross, K., *Computer Networking, A Top-Down Approach*, 4th Edition, Addison-Wesley, 2008.

[LAU02] Laude, J-P., *DWDM Fundamentals, Components and Applications*, Artech House, 2002.

[LIT02] Littman, M. K., *Building Broadband Networks*, CRC Press, 2002.

[MCC07] McCabe, J., *Network Analysis, Architecture and Design*, 3rd Edition, The Morgan Kaufmann Series in Networking, 2007.

[MOR04] Morais, D., *Fixed Broadband Wireless Communications: Principles and Practical Applications*, Prentice Hall, 2004.

[MUK06] Mukherjee, B., *Optical DWDM Networks*, Springer, 2006.

[ODR08] O'Driscoll, G., *Next Generation IPTV Services and Technologies*, Wiley, 2008.

[OLI06] Olifer, N., *Computer Networks: Principles, Technologies and Protocols for Network Design*, John Wiley & Sons, 2006.

[OPP04] Oppenheimer, P., *Top-Down Network Design*, 2nd Edition, CRC Press, 2004.

[OSB03] Osborne, E., and Simha, A., *Traffic Engineering with MPLS*, Cisco Press, 2003.

[PAP07] Papadimitriou, G., Papazoglou, C., and Pomportsis, A., *Optical Switching*, John Wiley & Sons, 2007.

[PER99] Perlman, R., *Interconnections: Bridges, Routers, Switches, and Internetworking Protocols*, 2nd Edition, Addison-Wesley Professional Computing Series, 1999.

[PIO04] Pioro, M., and Medhi, D., *Routing, Flow and Capacity Design in Communication and Computer Networks*, The Morgan Kaufmann Series in Networking, 2004.

[POI06] Poikselka, M., Niemi, A., Khartabil, H., and Mayer, G., *The IMS: IP Multimedia Concepts and Services*, Wiley, 2006.

[RIC02] Richardson, I., *Video Codec Design*, Wiley, 2002.

[ROS01] Rosen, E., Viswanathan, A., and Callon, R., Multiprotocol label switching architecture, *RFC 3031*, January, 2001.

[SCH06] Schmidt, K., *High Availability and Disaster Recovery: Concepts, Design and Implementation*, Springer, 2006.

[SHO01] Shooman, M., *Reliability of Computer Systems and Networks: Fault Tolerance, Analysis and Design*, Wiley, 2001.

[SIM06] Simpson, W., *Video Over IP, A Practical Guide to Technology and Applications*, Elsevier Focal Press, 2006.

[SPO02] Spohn, D., *Data Network Design*, Mc-Graw Hill, 2002.

[STE94] Stevens, W. R., *TCP/IP Illustrated, Vol. 1: The Protocols*, Addison-Wesley Professional Computing Series, 1994.

[STE00] Stern, T., and Bala, K., *Multiwavelength Optical Networks, A Layered Approach*, Addison-Wesley, 2000.

[SUM99] Summers, C., *ADSL Standards, Implementation and Architecture*, CRC Press, 1999.

[SZI04] Szigeti, T., and Hattingh, C., *End-to-End QoS Network Design: Quality of Service in LANs, WANs and VPNs*, Cisco Press, 2004.

[TAN04] Tan, N-K., *MPLS for Metropolitan Area Networks*, Auerbach Publications, 2004.

[THO99] Thomas, T., and Khan, A., *Network Design and Case Studies*, 2nd Edition, Cisco Press, 1999.

[WAN01] Wang, Z., *Internet QoS. Architecture and Mechanism for Quality of Service*, Morgan Kaufmann Publishers, 2001.

[YAM05] Yamanaka, N., Shimoto, K., and Oki, E., *GMPLS Technologies: Broadband Backbone Networks and Systems*, CRC Press, 2005.

[ZHA07] Zhang, Y., and Chen, H-H., *Mobile WiMax: Toward Broadband Wireless Metropolitan Area Networks*, Auerbach Publications, 2007.

[ZHU02] Zhu, Q., *High-Speed Clock Network Design*, Kluwer Academic Publishers, 2006.

Index